A Review of the New Initiatives at the NASA Ames Research Center

Summary of a Workshop

CHARLES W. WESSNER, EDITOR

Board on Science, Technology, and Economic Policy

Policy and Global Affairs

National Research Council

NATIONAL ACADEMY PRESS
Washington, D.C.

NATIONAL ACADEMY PRESS • 2101 Constitution Avenue, N.W. • Washington, D.C. 20418

NOTICE: The project that is the subject of this report was approved by the Governing Board of the National Research Council, whose members are drawn from the councils of the National Academy of Sciences, the National Academy of Engineering, and the Institute of Medicine. The members of the committee responsible for the report were chosen for their special competences and with regard for appropriate balance.

This study was supported by Contract No. NASW-99037-Task 103 between the National Academy of Sciences and the National Aeronautics and Space Administration. Any opinions, findings, conclusions, or recommendations expressed in this publication are those of the author(s) and do not necessarily reflect the views of the organizations or agencies that provided support for the project.

International Standard Book Number 0-309-07409-6

Limited copies are available from Board on Science, Technology, and Economic Policy, National Research Council, 1055 Thomas Jefferson Street, N.W., Suite 2014, Washington, D.C. 20007; 202-334-2200.

Additional copies of this report are available from National Academy Press, 2101 Constitution Avenue, N.W., Lockbox 285, Washington, D.C. 20055; (800) 624-6242 or (202) 334-3313 (in the Washington metropolitan area); Internet, http://www.nap.edu

Printed in the United States of America
Copyright 2001 by the National Academy of Sciences. All rights reserved.

THE NATIONAL ACADEMIES

National Academy of Sciences
National Academy of Engineering
Institute of Medicine
National Research Council

The **National Academy of Sciences** is a private, nonprofit, self-perpetuating society of distinguished scholars engaged in scientific and engineering research, dedicated to the furtherance of science and technology and to their use for the general welfare. Upon the authority of the charter granted to it by the Congress in 1863, the Academy has a mandate that requires it to advise the federal government on scientific and technical matters. Dr. Bruce M. Alberts is president of the National Academy of Sciences.

The **National Academy of Engineering** was established in 1964, under the charter of the National Academy of Sciences, as a parallel organization of outstanding engineers. It is autonomous in its administration and in the selection of its members, sharing with the National Academy of Sciences the responsibility for advising the federal government. The National Academy of Engineering also sponsors engineering programs aimed at meeting national needs, encourages education and research, and recognizes the superior achievements of engineers. Dr. William A. Wulf is president of the National Academy of Engineering.

The **Institute of Medicine** was established in 1970 by the National Academy of Sciences to secure the services of eminent members of appropriate professions in the examination of policy matters pertaining to the health of the public. The Institute acts under the responsibility given to the National Academy of Sciences by its congressional charter to be an adviser to the federal government and, upon its own initiative, to identify issues of medical care, research, and education. Dr. Kenneth I. Shine is president of the Institute of Medicine.

The **National Research Council** was organized by the National Academy of Sciences in 1916 to associate the broad community of science and technology with the Academy's purposes of furthering knowledge and advising the federal government. Functioning in accordance with general policies determined by the Academy, the Council has become the principal operating agency of both the National Academy of Sciences and the National Academy of Engineering in providing services to the government, the public, and the scientific and engineering communities. The Council is administered jointly by both Academies and the Institute of Medicine. Dr. Bruce M. Alberts and Dr. William A. Wulf are chairman and vice chairman, respectively, of the National Research Council.

Steering Committee for Government-Industry Partnerships for the Development of New Technologies*

Gordon Moore, *Chair*
Chairman Emeritus
Intel Corporation

M. Kathy Behrens
Managing Partner
Robertson Stephens Investment
    Management
and STEP Board

Michael Borrus
Managing Director
Petkevich & Partners, LLP

Iain M. Cockburn
Professor of Finance and Economics
Boston University

Kenneth Flamm
Dean Rusk Chair
    in International Affairs
LBJ School of Public Affairs
University of Texas at Austin

James F. Gibbons
Professor of Engineering
Stanford University

W. Clark McFadden
Partner
Dewey Ballantine

Burton J. McMurtry
General Partner
Technology Venture Investors

William J. Spencer, *Vice-Chair*
Chairman Emeritus
SEMATECH
and STEP Board

Mark B. Myers
Senior Vice-President, *retired*
Xerox Corporation
and STEP Board

Richard Nelson
George Blumenthal Professor of
    International and Public Affairs
Columbia University

Edward E. Penhoet
Dean, School of Public Health
University of California at Berkeley
and STEP Board

Charles Trimble
Vice-Chairman
Trimble Navigation

John P. Walker
Chairman and Chief Executive Officer
Axys Pharmaceuticals, Inc.

Patrick Windham
President, Windham Consulting;
    and Lecturer, Stanford University

---

*As of February 2001.

Project Staff*

Charles W. Wessner
Study Director

Alan H. Anderson
Consultant

McAlister T. Clabaugh
Program Associate

David E. Dierksheide
Program Associate

Contributors**

David B. Audretsch
Ameritech Chair of Economic
  Development
Director, Institute for Development
  Strategies
Indiana University

Michael I. Luger
Professor of Public Policy Analysis,
  Planning, and Business
University of North Carolina
  at Chapel Hill

---

*As of February 2001.
**Biographies of the contributors are included in Annex B.

For the National Research Council (NRC), this project was overseen by the Board on Science, Technology and Economic Policy (STEP), a standing board of the NRC established by the National Academies of Sciences and Engineering and the Institute of Medicine in 1991. The mandate of the STEP Board is to integrate understanding of scientific, technological, and economic elements in the formulation of national policies to promote the economic well-being of the United States. A distinctive characteristic of STEP's approach is its frequent interactions with public and private-sector decision makers. STEP bridges the disciplines of business management, engineering, economics, and the social sciences to bring diverse expertise to bear on pressing public policy questions. The members of the STEP Board* and the NRC staff are listed below:

Dale Jorgenson, *Chair*
Frederic Eaton Abbe Professor
    of Economics
Harvard University

M. Kathy Behrens
Managing Partner
Robertson Stephens Investment
    Management

Vinton G. Cerf
Senior Vice-President
WorldCom

Bronwyn Hall
Professor of Economics
University of California at Berkeley

James Heckman
Henry Schultz Distinguished Service
    Professor of Economics
University of Chicago

Ralph Landau
Consulting Professor of Economics
Stanford University

Richard Levin
President
Yale University

William J. Spencer, *Vice-Chair*
Chairman Emeritus
SEMATECH

David T. Morgenthaler
Founding Partner
Morgenthaler

Mark B. Myers
Senior Vice-President, *retired*
Xerox Corporation

Roger Noll
Morris M. Doyle Centennial
    Professor of Economics
Stanford University

Edward E. Penhoet
Dean, School of Public Health
University of California at Berkeley

William Raduchel
Chief Technology Officer
AOL Time Warner

Alan Wm. Wolff
Managing Partner
Dewey Ballantine

*As of February 2001.

STEP Staff*

Stephen A. Merrill
Executive Director

Philip Aspden
Senior Program Officer

Camille M. Collett
Program Associate

David E. Dierksheide
Program Associate

Charles W. Wessner
Program Director

Craig M. Schultz
Research Associate

McAlister T. Clabaugh
Program Associate

---

*As of February 2001.

**National Research Council
Board on Science, Technology, and Economic Policy**

Sponsors

The National Research Council gratefully acknowledges the support of the following sponsors:

National Aeronautics and Space Administration

Office of the Director, Defense Research & Engineering

National Science Foundation

U.S. Department of Energy

Office of Naval Research

National Institutes of Health

National Institute of Standards and Technology

Sandia National Laboratories

Electric Power Research Institute

International Business Machines

Kulicke and Soffa Industries

Merck and Company

Milliken Industries

Motorola

Nortel

Proctor and Gamble

Silicon Valley Group, Incorporated

Advanced Micro Devices

Any opinions, findings, conclusions, or recommendations expressed in this publication are those of the authors and do not necessarily reflect the views of the project sponsors.

Contents

FOREWORD	1
I. PREFACE	5
II. OVERVIEW AND SUMMARY OF THE WORKSHOP	15
III. INTRODUCTION	23
IV. PROCEEDINGS	
Welcome	35
Henry McDonald, Ames Research Center	
Opening Remarks	36
Zoe Lofgren, U.S. House of Representatives	
Panel I: A Technology Vision for NASA	38
Moderator: Edward Penhoet, University of California at Berkeley and Chiron Corporation	
NASA's Technology Strategy	38
Sam Venneri, NASA	
Ames' Technology Strategy	43
Henry McDonald, Ames Research Center	

Panel II: Research Parks: Concept, History, and Metrics   47
Moderator: David B. Audretsch, Indiana University

Presenter: Michael I. Luger, University of North Carolina   47
   at Chapel Hill

Discussant: Susan Hackwood, California Council on Science   53
   and Technology and UC Riverside

Panel III: The Ames Research Park: Goals and Metrics   57
Moderator: Patrick Windham, Stanford University
   and Windham Consulting

The Ames Strategic Plan   57
William Berry, Ames Research Center

Partnering with The University of California at Santa Cruz   64
M.R.C. Greenwood, University of California at Santa Cruz

The Role of Lockheed Martin   66
William Ballhaus, Lockheed Martin Corporation

The Role of Carnegie Mellon   69
Duane Adams, Carnegie Mellon University
James Morris, Carnegie Mellon University

Discussants: 74
   Robert Wilson, University of Texas at Austin
   Edward Penhoet, University of California at Berkeley
     and Chiron Corporation

Panel IV: SBIR Initiatives and Mission Objectives   78
Moderator: Burton McMurtry, Technology Venture Investors

In-Q-Tel: A "Nonprofit Venture Capital Fund"   78
Gilman G. Louie, In-Q-Tel

An "Enterprise Fund" for NASA   81
Robert L. Norwood, NASA

A Venture Capital Perspective on Research Parks   83
Kathy Behrens, Robertson Stephens Investment Management

Panel V: Ames as an Entrepreneurial Center: Opportunities and Challenges — 85
Moderator: Mark Myers, Xerox Corporation

Commercializing Technology — 85
Carolina Blake, Ames Research Center

The Experience of One Start-up Company — 87
Elizabeth Downing, 3D Technology Laboratories

Discussant: Jim Turner, House Science Committee — 89

Concluding Remarks — 92
Henry McDonald, Ames Research Center

Boxes within the Summary Report
Box A. Benefits of High-Technology Industries — 9
Box B. The Notion of Success — 28
Box C. Potential Risks and Guidance for Parks — 30
Box D. Regional Advantage in a Global Economy — 31

V. RESEARCH PAPERS

Science and Technology Parks at the Millennium: Concept, History, and Metrics — 95
Michael I. Luger, University of North Carolina at Chapel Hill

The Prospects for a Technology Park at Ames: A New Economy Model for Industry-Government Partnership? — 112
David B. Audretsch, Indiana University

VI. ANNEX

A. **Ames White Paper on the Research Park** — 137

B. **Biographies of Contributors** — 141

C. **Participants List** — 143

D. **Bibliography** — 147

Foreword

The National Aeronautics and Space Administration asked the Board on Science, Technology, and Economic Policy (STEP) to hold a one-day symposium to review the NASA Ames Research Center's plans to develop a science and technology park. As currently envisaged, the park will include three main elements: cooperative activities with two major universities (the University of California at Santa Cruz and Carnegie Mellon University) for a variety of educational missions; collaborative research with major high-technology industries in close proximity to Ames; and innovative efforts to encourage small business development. The workshop, held on 14 April 2000, brought together a Member of Congress, congressional staff, executive branch officials, representatives from the private sector, university officials, and regional economists to discuss the NASA Ames initiatives. In addition, two papers were commissioned, one to provide an analysis of the development and evaluation of S&T parks and another to review the unique features of the Ames proposal. The Ames S&T park will be an integral part of the 2,000-acre NASA Ames Research Center, located in Moffett Field, California. A description of the park concept, prepared by the NASA Ames Research Center, is included in the report in Annex A.

I
PREFACE

Preface

The technology-driven growth that has characterized the U.S. economy over the last decade has reinforced Americans' belief in the value of science and technology. New technologies are understood to be sources of strength for the economy as well as means of addressing national objectives, such as improved health care, a cleaner environment, and the exploration of space. Though less well understood, the government has long played an important role in stimulating scientific and technological advances, and this role has become increasingly important as we begin the new century.

The federal role is as diverse as it is important. The government directly stimulates scientific and technological research through its support of the large federal research agencies, such as the Department of Defense, the Department of Energy, the National Science Foundation, and the National Aeronautics and Space Administration (NASA). Much of this effort goes directly to universities, but some serves as a direct stimulus, translated through various mechanisms and programs, for private-sector activities that directly benefit the national economy and our capacity to achieve national goals.[1]

---

[1] This report is the second in the *Government-Industry Partnerships* series to focus on industry collaboration with national laboratories. The first analysis of these cooperative efforts focused on industry-laboratory partnerships at the Sandia National Laboratories. See National Research Council, *Industry-Laboratory Partnerships: A Review of the Sandia Science and Technology Park Initiative*, Charles W. Wessner, ed. Washington, D.C.: National Academy Press, 1999. The preface of that workshop report provides background information on the policy context of industry-laboratory collaboration, which is relevant to the Ames Research Center initiatives.

In recent years, adjustments in federal spending patterns have resulted in smaller research budgets for some federal agencies. These agencies have been challenged to meet and even extend mission objectives in the face of tighter budgets. In the case of NASA, agency planners have sought to reach their objectives through a "better, faster, cheaper" strategy that includes simplification, reliability, and versatility.[2] In addition to continuing its exploration of space, the federal space agency has also sought effective mechanisms to transfer its rich technological output into innovations of value in the commercial marketplace and to leverage its physical and human resources in new ways.[3]

The Ames Research Center's plans to develop a science and technology park represent a significant new initiative for NASA. This ambitious undertaking includes three main elements. Current plans call for cooperative arrangements with two major universities for a variety of educational missions, ranging from educational outreach to post-doctoral research. A central element of the initiative is to address common research goals through close cooperation between Ames and leading high-technology companies. In addition, the park will include a substantial emphasis on small business development, through the Ames incubator, NASA SBIR grants, and new approaches to funding for companies with technologies relevant to the NASA mission. These interrelated objectives and the proximity of the Ames Research Center to the technological ferment of Silicon Valley make this a unique chapter in NASA's continued efforts to leverage its resources.

THE ROLE OF STEP

The National Research Council's Board on Science, Technology, and Economic Policy (STEP) was founded in 1991 to improve understanding of the interconnections between science, technology, and economic policy and their importance to the American economy. The Board's activities have corresponded with increased recognition of the importance of support for basic, applied, and developmental research to continued economic growth.[4]

STEP recognizes that of the major investors in R&D—the federal government and private industry—the federal government has the primary but not exclusive responsibility to provide support for basic research. The government's role is central for at least four reasons: First, the federal government has the capacity to take a long view of research and provide the "patient funding" needed

---

[2] The NASA budget declined for several years; however, for fiscal year 2001 the NASA budget was increased to $14.285 billion, $633 million more than the fiscal year 2000 level.

[3] For example, NASA supports a substantial SBIR program totalling approximately $92.1 million in fiscal year 2000.

[4] For an informative discussion of different elements of the research process, see Donald Stokes, *Pasteur's Quadrant: Basic Science and Technological Innovation,* Washington, D.C.: Brookings Institution Press, 1997.

PREFACE

to put in place the foundations of the next generation of discoveries.[5] Second, the federal government is also uniquely placed to support the institutional framework of universities and laboratories to train researchers and develop new principles and processes, which ultimately contribute to scientific and economic progress. Third, the federal government is the primary entity with sufficient resources to make substantial, long-term, inherently uncertain investments in research and development of new technologies. Lastly, federal support for applied research and development also serves the public interest directly by increasing the government's capacity to achieve national missions in areas as diverse as health, public safety, and conservation.

These investments require a long-term view because the outcomes of basic scientific and technological research are inherently unpredictable. Basic insights that may seem useless in a practical sense often turn out to have immensely valuable applications years or decades after their discovery.[6] At the same time, many recognize that the taxpayers have a justified interest in seeing concrete economic and social benefits from this substantial public investment in R&D.[7]

Indeed, an important premise for economic policy, developed from the work of Robert Solow in the 1950s,[8] is that in the right circumstances, the outcomes of research stimulate commercially valuable innovations, and that these innovations

---

[5] As Richard Nelson notes, technological advance involves uncertainty in a fundamental way. The process is full of surprises, and it generally is not possible to predict the outcomes of research programs. R&D statistics and policy discussion often reflect assumptions of a linear model, by which innovation proceeds from fundamental discovery to applied research, and then to development and marketing. However, there is widespread recognition that this model is not adequate to describe the diverse origins and feedback loops of most real-world innovations. See Richard Nelson, "Technical advance and economic growth," in National Research Council, *Harnessing Science and Technology for America's Economic Future*, Washington, D.C.: National Academy Press, 1999 (www.nap.edu/html/harness_sci_tech/ch2.html). See also Branscomb's discussion of "basic technological research" in Lewis M. Branscomb and James H. Keller, eds., *Investing in Innovation: Creating a Research and Innovation Policy That Works,* Cambridge, MA: MIT Press, 1998, Chapter 5.

[6] Among many examples are the global positioning system (GPS), the popular navigational tool whose accuracy depends on the discovery by I. I. Rabi in the 1930s of magnetic resonance, which made possible the development of atomic clocks. Basic research played a similar role in optics. See the references to optics research in National Research Council, *Allocating Federal Funds for Science and Technology*. Washington, D.C.: National Academy Press, 1995, p. 77.

[7] For a discussion of the process of innovation and policies to stimulate it, see Branscomb and Keller, *Investing in Innovation, op. cit.,* Chap. 18. See also Lewis Branscomb, "The False Dichotomy: Scientific Creativity and Utility," *Issues in Science and Technology*, 16(1): 66, 1999. See also Lewis Branscomb, *Taking Technical Risks: How Innovators, Managers, and Investors Manage Risks in High-Tech Innovations.* Cambridge, MA: MIT Press, forthcoming, chapter 5.

[8] Solow found that a small fraction of economic growth could be assigned to labor, and that capital formation accounted for approximately one-third of growth. This leaves a large "Solow residual" that is assigned to technological progress, exogenously determined. More recently, new growth theory has emphasized technology as an "endogenous" factor. Endogenous growth theory postulates several channels through which technology, human capital, and the creation of new ideas enable a "virtuous

drive economies.[9] This premise underlies the rationale for modern federal investments in R&D, which ultimately serve the public interest in the form of improved products, processes, and understanding.[10]

A second premise, still unfolding, is that the most powerful fuel for the economy is found in research that underlies high-technology innovations—those that involve highly advanced or specialized systems or devices. The so-called "knowledge-based" fields of research, such as software engineering, wireless communications, biotechnology, and artificial intelligence, all grew out of basic research whose outcomes were unforeseeable. They may prove to be as dominant in the economy of the information age as oil and steel were in the economy of the industrial age.[11] According to a recent study by the Milken Institute, the growth of the high-technology sector since the 1990-91 recession has been four times as rapid as that of the aggregate economy.[12] This sector plays a major and disproportionate role in the national R&D effort and developing opportunities for economic growth and job creation.

---

circle" and feedback to economic growth. See Paul Romer, "Endogenous technological change," *Journal of Political Economy,* 1990, 98:71-102. This understanding is critical in attempting to determine the contribution of new technologies (such as information technology) to the growth process and, specifically, to the growth of productivity.

[9] The macro-economic environment greatly conditions the returns to these investments. For example, European policymakers have recently wrestled with the failure of a vibrant R&D enterprise to convert research into technological and commercial success. See European Commission, *Research and Technology: the Fourth Framework Programme (1994-1998)*, Brussels, Belgium, 1995, p. 12. The 1995 report cites three features of the European research system to partly explain these weaknesses: the inadequate translation of research results into commercial applications, insufficient investment in research and technology development programs in the fields of education and training, and the fragmentation and lack of coordination in European research efforts.

[10] The impact of such programs on international research cooperation and the multilateral trading system are of considerable interest not only to U.S. research agencies but to policymakers around the world. Reflecting this interest, these topics were taken up by STEP in conjunction with the Hamburg Institute for Economic Research and the Institute for World Economics in Kiel in a collaborative project. One of the principal recommendations of the joint report emerging from that study called for an analysis of the principles of effective cooperation in technological development. See National Research Council, *Conflict and Cooperation in National Competition for High-Technology Industry*, Washington, D.C.: National Academy Press, 1996.

[11] See Ross C. DeVol et al., *America's High-Tech Economy: Growth, Development, and Risks for Metropolitan Areas,* Santa Monica, CA: Milken Institute, July 13, 1999 (www.milken-inst.org). The report focuses on the value of output for industries that may be considered high-technology, including manufacturing industries (drugs, computers and equipment, communications equipment, and electronic components) and service industries (communications services, computer and data processing services, and research and testing services).

[12] *Ibid.*

> **Box A. Benefits of High-Technology Industries**
>
> High-technology industries bring special benefits to national economies. These industries are associated with innovation, which means they tend to gain market share, create new product markets, and use resources more productively than traditional industries. They also perform larger amounts of R&D (spending over 10 percent of revenues on research, vs. 3 percent for more traditional industries). This high level of expenditure also creates positive spillover effects that benefit other commercial sectors. A substantial economics literature underscores the high returns of technological innovation, with private innovators obtaining rates of return in the 20-30 percent range and spillover (or social return) averaging about 50 percent.
>
> There are also positive spillover effects to other commercial sectors through the generation of new products and processes that lead to productivity gains and new opportunities. For example the surging capabilities and falling costs of new technologies based on semiconductors have enabled new methods of manufacturing in steel, automobiles, and aerospace, and major advances in consumer electronics and even agriculture.
>
> Consequently, high-technology industry in many regions is seen as a major source of national economic growth in all of the major industrialized countries. In particular, high-technology firms are valued as creators of high value-added manufacturing and high-wage employment.
>
> —National Research Council, *Conflict and Cooperation*, 1996, p. 34.

PROJECT ORIGINS: EXAMINING PARTNERSHIPS

The growth in government programs to support high-technology industries raises new challenges and opportunities for NASA and the other research-intensive federal agencies. NASA is challenged to "do more with less" in the face of a declining budget and a strong desire on the part of the government to make the most productive use of its resources. At the same time, the space agency has opportunities to combine its technological assets with its considerable experience in creating partnerships with private firms to capitalize on the value of these assets. These activities have encouraged NASA, and other agencies, to explore new models for government-industry partnerships.

Reflecting the interest of policy makers in this topic, the STEP Board initiated the project on "Government-Industry Partnerships for the Development of New Technologies," which has benefited from broad support among federal agencies. These include the U.S. Department of Defense, the U.S. Department of Energy, the National Science Foundation, the National Institutes of Health, the National Cancer Institute, the National Institute of General Medical Sciences, the National Aeronautics and Space Administration, and the National Institute of Standards and Technology, as well as a diverse group of private corporations listed in the front of the report. To carry out this analysis, the STEP Board has assembled a distinguished multidisciplinary steering committee for government-industry partnerships, listed in the front of this report. The Committee's principal tasks are to provide overall direction and relevant expertise to assess the issues raised by the project. At the conclusion of the project, the Steering Committee is to develop a consensus report outlining their findings and recommendations.

As a basis for the consensus report, the Steering Committee is commissioning research and convening a series of fact-finding meetings in the form of workshops, symposia, and conferences as a means of both informing its deliberations and addressing current policy issues affecting government-industry partnerships. As the project progresses, the Steering Committee is making recommendations and findings on major elements of its work, particularly in response to requests from participating agencies. This report can therefore be seen as both an input into the broader Academy assessment of partnerships and as a contribution to national policy making.

ACKNOWLEDGEMENTS

A number of individuals deserve recognition for their contributions to the preparation of this report and for their willingness to serve as reviewers. On behalf of the STEP Board we would like to express special recognition to Henry McDonald, Director of Ames Research Center; William Berry, Deputy Director of Ames Research Center; Carolina Blake, Chief of the Commercial Technology Office, Ames Research Center; Robert Norwood, Director for Commercial Development and Technology Transfer, NASA Headquarters; and Diana Hoyt, Senior Policy Analyst, NASA Headquarters. Their interest and commitment to an objective assessment of the Ames S&T Park initiative was crucial to the success of this review. Similarly, special recognition is due to David Audretsch, Ameritech Chair of Economic Development and Director of the Institute for Development Strategies, Indiana University; and Michael Luger, Professor of Public Policy Analysis, Planning, and Business and Founding Director of the Office of Economic Development at the University of North Carolina at Chapel Hill; for their many valuable insights. We also wish to thank Alan Anderson for his role in the preparation of the draft manuscript for this volume. Among the STEP staff, special recognition goes to David Dierksheide and McAlister Clabaugh for their support of the meetings at NASA Ames Research Center and their care in preparing and editing the manuscript for publication. Their enthusiasm and interest were essential for STEP to meet NASA's request for a review of the Ames initiatives.

This report has been reviewed in draft form by individuals chosen for their diverse perspectives and technical expertise, in accordance with procedures approved by the NRC's Report Review Committee. The purpose of this independent review is to provide candid and critical comments that will assist the institution in making its published report as sound as possible and to ensure that the report meets institutional standards for objectivity, evidence, and responsiveness to the study charge. The review comments and draft manuscript remain confidential to protect the integrity of the deliberative process. We wish to thank the following individuals for their review of this report: Dr. Kathryn L. Combs, University of St. Thomas, St. Louis; Dr. Albert N. Link, University of North Carolina at Greensboro; Dr. Edward J. Malecki, University of Florida; Mr. John C. McDonald, MBX, Inc.; Mr. Thomas F. Widmer, Thermoelectron, *retired*; and Dr. Robert H. Wilson, University of Texas at Austin. Although the reviewers listed above have provided many constructive comments and suggestions, they were not asked to endorse conclusions or recommendations nor did they see the final draft of the report before its release. The review of this report was overseen by Alexander H. Flax of the Washington Advisory Group. Appointed by the National Research Council, he was responsible for making certain that an independent examination of this report was carried out in accordance with institutional procedures and that all review comments were carefully considered. Responsibility for the final content of this report rests entirely with the authoring committee and the institution.

Charles W. Wessner

II

OVERVIEW and SUMMARY OF THE WORKSHOP

Overview and Summary

OVERVIEW

In order to perform its unique missions, NASA is seeking to capitalize on its existing assets and promising new technological trends in biotechnology, nanotechnology and information technology. As an integral part of the NASA infrastructure, the Ames Research Center, at Moffett Field, California, has developed a strategic plan to make use of its extensive human and physical resources in ways that are both consistent with NASA's overall mission goals and which are effective at leveraging its own particular research capabilities and exceptional location in the heart of Silicon Valley.[1]

The Ames Research Center is embarking on a program to develop a science and technology park bringing together leading high-technology companies and universities, such as the University of California at Santa Cruz and Carnegie Mellon, to contribute to Ames' exceptional mission and to the educational and research requirements of this unique American cluster of economic growth and invention. The park is to include shared research facilities and public-private cooperation in teaching and training with the goal of contributing to NASA's core missions of research, exploration, and discovery. An additional objective is to facilitate NASA's increased emphasis on commercializing technologies developed by agency scientists and engineers and contribute related national benefits such as higher computer dependability. Other initiatives under consideration

---

[1] For an overview of the Ames proposal, see the White Paper submitted by NASA in Annex A. For additional information on the concept, see the presentations by NASA's Sam Venneri in Panel I, William Berry in Panel III, and Robert Norwood in Panel IV.

include the integration of SBIR grants with a planned on-site incubator, virtual or distance collaboration, and possibly a new public venture capital program.

Given the scope and ambition of these objectives, the NASA Administrator, Daniel Goldin, asked the NRC's Board on Science, Technology and Economic Policy to review the Ames initiatives. The STEP Board, through its Chairman, Dale Jorgenson, and Vice-Chairman, Bill Spencer, accepted the NASA request and, after a series of preliminary meetings, convened a one-day workshop on 14 April 2000 at Ames Research Center. Although there was a broad range of issues to consider in a single workshop, the discussion did succeed in raising many of the issues that Ames might expect to encounter as it proceeds with its plan to "invent" the Ames Research Park.

Workshop participants raised and debated a variety of issues affecting the management, operation, focus, and metrics of the proposed park:

- the advantages for Ames and NASA of participating in new and emerging technologies;
- the need for private sector participation to share costs, risks, and expertise;
- the potential gains from leveraging the assets of Ames to advance NASA missions;
- the potential contribution of expanded educational facilities to meeting the pressing need for graduate and postgraduate training and research;
- the challenge of addressing effectively multiple and sometimes competing objectives;
- the local challenges to development, including a tight labor supply, high housing costs relative to the rest of the nation, and growing environmental constraints;[2]
- the opportunity for NASA and its partners to more fully capture the potential of current and future R&D investments; modified by
- the inherent complexity of public-private technological transfer, especially for fast-paced commercial applications, compared with longer-term NASA mission-oriented research.

Outside Analysis

To complement the Board's discussion of the NASA proposal, two papers were commissioned as part of the preparation of the report. The paper by Michael I. Luger, *Science and Technology Parks at the Millennium: Concept, History, and Metrics*, provides a comprehensive overview of the science and technology and related park developments around the world in order to give NASA a broader

---

[2] While these issues are unquestionably relevant to the Ames initiative, the focus of the workshop was primarily on issues of national policy where the Board has substantial expertise.

context for its planning activities. The commissioned work by David B. Audretsch, *The Prospects for a Technology Park at Ames: A New Economy Model for Industry-Government Partnership?*, underscores the unique features of the Ames Research Park proposal. Rather than seeking to provide an engine of growth for the region via outward technology transfer, Audretsch observes that the goal of the Ames park is to enable NASA to achieve its mission by providing economical access to technological capabilities external to NASA. This would occur both through the inward transfer of technologies developed outside of NASA and through the joint development of new technologies by NASA in conjunction with its partners in private industry and the universities. Audretsch also proposes a series of metrics to monitor and measure the impact of the Ames Research Park.

As both papers affirm, the creation of a successful S&T park requires effective cooperation by many parties. Both the participants in the discussion and the commissioned analysis highlighted that the management challenge for Ames will be to accomplish the multiple objectives of this initiative, in collaboration with the industry and university partners as well as the state and local governments, while keeping in mind the need for clear goals and appropriate metrics to measure progress in this innovative undertaking.

SUMMARY OF THE WORKSHOP

Representative Zoe Lofgren, who serves on the Space Subcommittee of the House Science Committee, welcomed the participants to her district in Silicon Valley and underscored the importance of federal funding both to advance the nation's research agenda and to educate the next generation of scientists and engineers.

The workshop itself was divided into five main panels for presentations and discussion.

Strategic Direction

The first panel dealt with NASA's technology strategy, which will focus on three primary theme areas: nanotechnology, biotechnology, and information technology. The strategy is to integrate these three systems in the agency's search for evolvable, adaptable, extremely tough, self-repairing systems. Ideally, such systems would be able to create information and knowledge from data, perform self-diagnosis and repair, and make decisions – in effect, to "think for themselves." Ames plans to pursue this strategy through research partnerships with both private firms and universities.

Ames Advantages

The second panel, led by Michael Luger of the University of North Carolina at Chapel Hill, discussed the concept, history, and metrics of research parks.

Dr. Luger described the difficulty of predicting the success of new parks and the absence of uniform standards by which to measure success. He cautioned that most parks do not generate tangible benefits that exceed their costs, and even those that are successful require long incubation periods. However, Dr. Luger also enumerated advantages an Ames research park would have, including pre-existing intellectual prominence, available and essentially cost-free real estate, access to the exceptional technological and financial resources of Silicon Valley, a historic relationship with Lockheed Martin, the support of the local communities, and the considerable intellectual and institutional resources of its academic partners, the University of California and Carnegie Mellon University. These are substantial assets and, as Professor Audretsch's paper argues, they distinguish the Ames initiatives from more traditional S&T parks.[3]

Public-Private R&D Partnerships

The third panel described in more detail the goals and metrics for the park. Dr. William Berry, Ames' Deputy Director, described the primary objective to extend and deepen the R&D capabilities of Ames through R&D partnerships that would focus primarily on the goals of information technology, nanotechnology, and biotechnology. Partners would create or renovate their own facilities on Ames property. In return for land and a unique relationship with NASA researchers, those partners would reinvest funds gained through partnership activities in collaborative activities at Ames. Although there would be no direct revenue stream for Ames, partnerships would be fueled primarily by nonappropriated funds.

Private Management

The discussion of goals and metrics was continued with a description by William Ballhaus of Lockheed Martin's role as partner. Lockheed Martin, which has collaborated with Ames on the strategic plan, already performs virtually all of its research in the context of government and industry partnerships, and Ames would benefit from this experience. At Ames, Lockheed would direct a Research Initiative Fund to support new programs, consult on human resources and regional development programs, and direct most of the large structural improvements to the facility.

Expanded Research and Educational Outreach

Dr. M.R.C. Greenwood, chancellor of the University of California at Santa Cruz, described her institution's primary objectives in forming a partnership with

---

[3] See the papers by Michael Luger and David Audretsch in this volume.

Ames, emphasizing both the importance of common research objectives and the opportunity to meet pressing educational needs of the state. One new area of collaborative research, for example, will involve Santa Cruz taking the lead role in a new astrobiology facility. Collaboration will also enable Santa Cruz to expand its programs of educational outreach, K-12 teacher training, and technical retraining.

Partners in Education and Research

Drs. Duane Adams and James Morris from Carnegie Mellon described their university's goals in forming partnerships with Ames, including collaborative research in robotics, information technology, software engineering, human-computer systems, and "dependable computing," involving both students and faculty. Moderator Edward Penhoet emphasized the difficulty of forming truly collaborative partnerships, in the sense of sharing insights, diversity of expertise, and leadership.

An Innovative Investment Model

On panel four, Gilman Louie, who directs an investment firm called In-Q-Tel on behalf of the Central Intelligence Agency, described the CIA's effort to form equity-based partnerships, rather than contractual relationships, with high-tech firms. In-Q-Tel believes that equity partnerships have several advantages over contracts. First, traditional contracts contain goals and metrics that must be known at the outset; with new technologies, it is seldom possible to know the outcome of a project in advance. Second, part of a traditional contractor's motive to succeed is to avoid penalties built into the contract. With an equity partnership, the partner is motivated by the hope of profiting from the success of a new technology. This more positive-sum approach was suggested as better suited to the requirements of rapidly changing technologies.

An Enterprise Fund

Taking up this theme, Robert Norwood of NASA described the prospect of a technology investment fund (an "enterprise fund") for the space agency with the goals of 1) identifying NASA technologies with strong commercial potential, and 2) finding corporate partners capable of commercializing those technologies. The fund would operate as a nongovernmental entity, free of government rules and constraints. It would combine the research strength of NASA, which reduces technological risk, with the business strength of the investment community, which reduces business risk.

Kathy Behrens, an investment banker, pointed out that the differences between the cultures and goals of government agencies and those of the private investment world might bring difficulties to such partnerships. She cautioned that agencies should avoid the difficult job of "trying to find a home for our technologies."

Encouraging Entrepreneurial Activity

On panel five, Carolina Blake of the Ames Research Center discussed entrepreneurial activity, beginning with the Commercial Technology Office at Ames, whose objectives include technology assessment, marketing, licensing intellectual property, and recruiting partners to work on technologies that are both critical to Ames' mission and potentially profitable for the companies.

Elizabeth Downing, head of 3D Technology Laboratories, described the funding difficulties of her start-up company. Neither venture funds nor technology firms were interested in funding a technology that requires time to develop. She said that without government support in the form of NSF, DARPA, and DoD SBIR grants, as well as a recent ATP award, it would have been impossible to make progress with her promising technology.

Cautious Encouragement

In conclusion, Jim Turner of the House Science Committee praised the effort at Ames as an innovative use of the space program's resources. Turner also offered a word of caution, advising that Ames take special care to avoid the perception of "corporate favors." Charles Wessner suggested that the flexibility of the Space Act, and its legitimacy, should be drawn on as the project goes forward. While the Space Act authority allows partnering, he cautioned that the technological and perhaps political risk associated with equity investments or venture activities should be kept in mind. Success rates, even for outstanding venture capital firms, are not always high enough to meet Washington's admittedly ill-defined standards. Nonetheless, he suggested that it is only fair to observe that, taken as a whole, this ambitious cooperative initiative does address needs central to the NASA mission and may provide a means of meeting educational needs which are equally central to the continued development of the region.

As the reader can appreciate, this was a broad range of issues to consider in a single symposium. A principal goal of the workshop was to raise many of the issues that Ames might expect to encounter as it moves forward with its plan to create a center of instruction and collaborative research contributing to the NASA mission and the needs of the nation.

III
INTRODUCTION

Introduction

For the federal government, a significant task of the new century is realigning the missions of its major science and technology laboratories. As the nation's attention has turned from the tensions of the Cold War to an intensely competitive global environment, federal agencies are challenged to find productive new ways to utilize their highly skilled human resources and their extensive—sometimes unique—physical assets.

This task is changing the strategies of federal agencies in fundamental ways. Within NASA, for example, the space agency's mission now includes an increased emphasis on commercializing technologies that are developed by agency scientists and engineers. And the development of tomorrow's technologies has shifted, in the words of NASA's chief scientist, toward "highly complex, first-of-a-kind missions which cannot be accomplished or afforded using current systems."[1]

To execute that shift, the agency plans to realign its own research mix by focusing on three cutting-edge areas—biotechnology, nanotechnology, and information technology. Its strategic challenge is to integrate those research areas into a new "mission triangle" for the 21st century.

A NEW STRATEGIC PLAN FOR AMES

As a major participant in this mission, the Ames Research Center, at Moffett Field, California, has undertaken a major effort to develop its own strategic plan.[2]

---

[1] See the remarks of Dr. Samuel Venneri, NASA's Chief Technologist, in the Proceedings, Panel I.
[2] For an overview of the Ames proposal, see the White Paper submitted by NASA in Annex A. For additional information on the park concept, see the presentations by NASA's Sam Venneri in Panel I, William Berry in Panel III, and Robert Norwood in Panel IV.

Ames' objective is to make use of its extensive human and physical resources in ways that are both consistent with NASA's overall mission goals and effective at leveraging its own particular research capabilities and exceptional location.

The crux of Ames' strategic plan is the formation of a "research park"[3] at the Ames facility that will feature a network of partnerships with private research firms and major universities. Ames brings to that plan its own substantial expertise in each of the three mission triangle fields, as well as 2,000 acres of land that includes undeveloped space and unused structures. Not the least of the park's assets is its location in the heart of Silicon Valley.

In recent years, federal agencies have tried various strategies to launch their technologies into the marketplace, with varying success. The plan to create onsite partnerships is interesting for several reasons. Most obviously, it binds NASA directly and proximately with private firms that have intimate knowledge of the high-technology markets.[4] Equally important, the plan avoids generally ineffective "technology push" techniques in favor of projects that will be driven by market needs. Ames' favorable location can facilitate gaining accurate knowledge of those needs.

Doing more within existing budgets has become the hallmark of NASA over the last decade. In keeping with this imperative, Ames intends to expand its mission and promote commercialization without significantly increased budgetary support. As described in the Proceedings by Center Director Henry McDonald and Deputy Director William Berry, new research efforts would be financed by both leasing revenues from partners and from the profits of partnership enterprises. At the same time, its existing Commercial Technology Office is prepared to assist in the recruitment of partners whose work is consistent with NASA's mission and to expand its existing mission of transferring technology to the private sector.

---

[3] "Research park" is the most common designation for an association of enterprises that focus on research and development. However, "research," "science," and "technology" park are used more or less interchangeably in the U.S. and Canada; the terms "science park" and "technopole" are more common in Europe and Asia. See the analysis of Michael Luger in this volume.

[4] As Linda Cohen and Roger Noll point out, one of the strengths of industry-led collaborations, such as Sematech, is the market savvy of participating firms. Linda R. Cohen and Roger G. Noll, *The Technology Pork Barrel,* Washington, D.C.: The Brookings Institution, 1991, chapter 12. Peter Grindley, David Mowery, and Brian Silverman make a similar point, noting that the consortium's goals changed over time, reflecting the changing perceptions of its members' needs. This operational flexibility is probably essential in an industry evolving as rapidly as the semiconductor industry. See Peter Grindley, David Mowery, and Brian Silverman, "SEMATECH and Collaborative Research: Lessons in the Design of High-Technology Consortia," *Journal of Policy Analysis and Management,* 13(4), 1996.

RESEARCH PARKS

Goals for research parks vary with local conditions, the goals of their organizers, and the assets and resources available to the enterprise. As Luger and Goldstein's research has made clear, the objectives of private research park developers are often not the same as those of public sector entities.[5] The concept of the research park is a modern phenomenon; most of them have been initiated in the last two decades. Research parks are defined by Luger and Goldstein as "organizational entities that sell or lease spatially contiguous land and/or buildings to businesses or other organizations whose principal activities are basic or applied research or development of new products or processes."[6] This concept evolved from the older industrial parks, whose first incarnation was the Central Manufacturing District of Chicago, established in 1905. The modern prototype of the research park appeared after World War II with the creation of Menlo Park, California, in 1948. The most successful parks to date were developed in the 1950s and 60s: Stanford Industrial Park in 1953, in northern California; Research Triangle Park in 1958, in central North Carolina; and Waltham Industrial Center in 1954, associated with other developments on Route 128 west of Boston.[7]

By 1980 some 20 parks had been formed—enough to attract the attention of regions across the country. During the next 10 years the number of parks increased five-fold. Since 1990, however, reductions in government and industry spending have slowed this pace of growth considerably. Between 1990 and 1995, the number of park formations dropped to 30, a more cautious pace that has continued to the present. Still, every state in the U.S. has at least one research park, and most have two or three.[8]

---

[5] Michael I. Luger and Harvey A. Goldstein, *Technology in the Garden: Research Parks & Regional Economic Development,* Chapel Hill: University of North Carolina Press, 1991 See also the paper by Michael Luger, "Science and Technology Parks at the Millennium" in this volume. Luger and Goldstein note, for example, that Research Triangle Park was initially a state effort, whose goals were to attract high-technology industry, and thereby create jobs and improve the perception of the state as a high-tech research center. The objectives of parks created or controlled by the private sector tend to put profit and occupancy ahead of more lofty developmental—but often unprofitable—goals. It is for this reason that many research parks go "down-market" from exclusively R&D activities to manufacturing, assembly, and distribution. Reflecting this evolution, Luger observes that one quarter of the parks he reviewed failed as real estate projects and one half of the remainder changed their focus to remain viable.

[6] Luger and Goldstein, *Technology in the Garden, op.cit.,* p. 5.

[7] Denise Drescher, *Research Parks in the United States: A Literature Review,* Department of City and Regional Planning, University of North Carolina at Chapel Hill, April 13, 1998, (www.unc.edu/~drescher/litrev.htm).

[8] These figures are from the Association of University-Related Research Parks, which was founded in 1986. See www.aurrp.org.

OBJECTIVES: JOBS, GROWTH, AND SYNERGIES

Overall, most "research parks" have been initiated by real-estate developers who saw the development of research capabilities as only one of several objectives. The three most common reasons to develop research parks are to create jobs, to raise a region's status by bringing in high-growth industries, and to create synergies between different firms. The country's largest park, Research Triangle Park (comprising some 5,000 acres), was established under Governor Luther Hodges for the purpose of creating jobs for the college-trained youth of North Carolina.[9]

A significant number of parks are collaborations or partnerships between the public and private sectors. These are usually promoted by regional or local governments, often in association with nearby universities and with private companies that occupy the resulting space.[10] Public development money is often involved, and the mission statements of many parks include active participation in technology transfer to the private sector and participation in the economic development of their cities, regions, and states.[11] Some parks, like University Heights Science Park in Newark, New Jersey, have strong urban renewal missions.[12] Some of these, like Kendall Square in Cambridge, Massachusetts, are tightly clustered around a single university (in this case, the Massachusetts Institute of Technology).[13]

AMES ADVANTAGES

NASA is following an approximate model that has become well established as a means of advancing the objectives of research and technology transfer. It is typical in being affiliated with one or more universities; it is unusual, though by no means unique, in locating on the site of a national laboratory. The Ames park is also unusual in several other respects: 1) It has the ability to construct its own campus on-site; 2) it is located adjacent to Silicon Valley, the world's largest concentration of high-tech firms and entrepreneurs;[14] and 3) it has, in advance of

---

[9] Drescher, *Research Parks in the United States, op. cit.,* p. 2.

[10] For a description of research parks from a real estate perspective, see Rachelle Levitt, ed., *The University/Real Estate Connection: Research Parks and Other Ventures,* Washington, D.C.: Urban Land Institute, 1987.

[11] Thomas W. Durso, "Home-Grown R&D," *The Scientist* 10[14]:1, July 8, 1996.

[12] Rachell Garbarine, "Newark's Science Park Takes Another Step Forward," *The New York Times on the Web,* Nov. 23, 1997.

[13] Some 70 biotech firms are located within five miles of Kendall Square, a recently renovated site of urban decay. Carey Goldberg, "Across the U.S., Universities Are Fueling High-Tech Economic Booms," *The New York Times on the Web,* Oct. 8, 1999.

[14] Ross C. DeVol et al., *America's High-Tech Economy: Growth, Development, and Risks for Metropolitan Areas,* Santa Monica, CA: Milken Institute, July 13, 1999 (www.milken-inst.org). DeVol

construction, already secured both academic and industrial partners of considerable resources and reputation.[15]

MANY GOALS—DIFFERENT METRICS

The success of research parks is difficult to measure. As suggested above, they vary a great deal in terms of objectives, types of facilities, and supporting institutions. Reflecting the many goals and different circumstances, there is a corresponding lack of agreement on metrics. There is little research or quantitative evidence that accounts for failed research parks.[16]

The comprehension of the research park phenomenon has also been blurred by political, ideological, and business biases, in that each participant in the development of a park has a particular definition of success.[17] Parks are often seen as cure-alls by developers and local governments, who may hope for job generation, income growth, greater income equality, expanded opportunities for certain groups within the labor force, and economic restructuring of the region.[18]

Park developers often describe beneficial changes as results of park creation, but the park's contribution to change is difficult to gauge. For example, many of the jobs attracted to a park might have come to the region in the absence of a park. Conversely, jobs created outside a park may be a function of the park's existence. Finally, costs as well as benefits must be examined to determine "success," including indirect expenditures on land acquisition and infrastructure development, tax expenditures from financial inducements used by government, and the opportunity cost of land used versus other types of uses.[19]

---

uses a measure of "high-tech spatial concentration" to describe "Tech-Poles" through the country. In his evaluation, the Tech-Pole of Silicon Valley ranks more than three times higher than its closest competitor. "As a Tech-Pole," DeVol writes, "the gravitational pull of the San Jose metro area, home to Hewlett-Packard, Applied Materials, Sun Microsystems, Intel, Cisco Systems, Oracle, and Silicon Graphics, is unparalleled." p. 6.

[15] Ames has planned university partnerships with Carnegie Mellon University and the University of California at Santa Cruz, which bring considerable strengths in science and technology, and an industrial partnership with Lockheed Martin (one division of which abuts the Ames property). It has also discussed potential partnerships with high-tech firms which have expressed interest in participating. See the presentation in this volume of M. R. C. Greenwood, Chancellor of the University of California at Santa Cruz.

[16] Drescher, *Research Parks in the United States, op. cit.*, p. 4.

[17] *Ibid.*, p. 4.

[18] Luger and Goldstein (*Technology in the Garden, op. cit.*, p. 34) write: "One of the conceptual difficulties is that there is no consensus about the definition of success.... The most commonly cited goals relate to economic development. But both the literature and our data from interviews with park developers, elected officials, university administrators, business leaders, and others confirm the existence of other goals, including technology transfer, land development, and enhancement of the research opportunities and capacities of affiliated universities."

[19] *Ibid*, p. 35.

To address the metrics gap, Luger and Goldstein have developed multiple approaches to measure success. One is to evaluate performance against stated goals, as written into legislation and found in documents and interviews. They also use a multiple case-study approach, along with a quasi-experimental design, comparing areas with parks to similar areas without parks. Overall, they report that about half of all parks do not succeed, and of the remainder, half shift their focus from research to become office, industrial, or mixed-use parks.[20]

Box B. The Notion of Success

"The overall policy lesson we have drawn from this analysis is that in many regions research parks by themselves will not be a wise investment. The success rate among all announced parks is relatively low. . . Research parks will be most successful in helping to stimulate economic development in regions that already are richly endowed with the resources that attract highly educated scientists and engineers."

—Luger and Goldstein, *Technology in the Garden*, p. 184.

For those parks that succeed, Luger and Goldstein and others conclude that the most notable consequence of developing a park is likely to be induced growth of R&D activity. R&D businesses are likely to cluster within the regions in order to share a specialized labor force, university facilities and expertise, business services, a certain type of social and cultural environment, and access to technical and market information. Consequently, "once a region 'takes off' with a successful research park, it should continue to experience growth in the R&D sector."[21]

CLUSTERS AND GROWTH: THE ROLE OF HIGH-TECHNOLOGY RESEARCH

The suggestion that parks can induce the growth of R&D activity is mirrored by evidence that high-tech activity stimulates regional growth. The Milken Institute study referred to above compared the economic success of areas with those areas' concentration of high-tech research. The authors concluded that high-tech activity explained 65 percent of the difference in economic growth among various metropolitan regions during the 1990s, and that "research centers and institutions are undisputedly the most important factor in incubating high-tech industries."[22] There is also evidence that ". . . high-technology firms are associated

---

[20] *Ibid*, p.1.
[21] *Ibid*, p. 22.
[22] DeVol, et al., *America's High-Tech Economy, op. cit.*, p. 5.

with innovation... and hence gain market share, create new product markets, and use resources more productively." Such firms tend to perform larger amounts of R&D than more traditional industries and create positive spillover effects that benefit other commercial sectors by generating new products and processes.[23]

RISKS

The process of creating a new research park carries substantial risks. Many (perhaps most) parks have yet to recruit the expected number of tenants, and there is no single formula by which one can avoid these risks. An official of the AUURP commented that many parks have made the error of assuming that big companies would move into new park spaces—just as IBM became the first tenant of Research Triangle Park. More often, a park has to "grow its own" tenants, a process that requires time.[24]

Certain risks are inherent to the R&D universe. High-tech tenants are influenced by business cycles that traditionally experience sharp swings. Such tenants are also vulnerable to cuts in federal or industry spending, especially because long-term research is often seen by budget planners as discretionary.[25]

ADVANTAGES OF CO-LOCATION

One feature that favors the success of parks and attracts tenants is the co-location of participants. Many firms, industries, and regions that are successful have formed collaborative relationships with other firms, agencies and universities to leverage the benefits of cooperation.[26] Such benefits already characterize the regional Silicon Valley network around Ames, which have been described as more "flexible, technologically dynamic, and tolerant of failure" than regions in which experimentation and learning are confined to individual firms.[27]

---

[23] The report adds that private innovators obtain a rate of return in the 20-30 percent range with the spillover (or social return) averaging about 50 percent. Positive spillovers are often locally concentrated. National Research Council, *Conflict and Cooperation in National Competition for High-Technology Industry*, Washington, D.C.: National Academy Press, 1996, pp. 33-35.

[24] Durso, "Home-Grown R&D," *op. cit.* p. 4. Also, Luger and Goldstein observe that it may take a decade or more for a park to mature or "succeed" (*Technology in the Garden, op. cit.,* p. 44).

[25] DeVol, et.al., *America's High-Tech Economy, op. cit.,* p. 9.

[26] Jane Fountain writes that social capital includes shared resources, shared staff and expertise, group problem-solving, multiple sources of learning, collaborative development, and diffusion of innovation, all of which are abundantly present in Silicon Valley. She suggests that social capital is as important as human and physical capitals. Jane Fountain, "Social Capital: A Key Enabler of Innovation," in Lewis M. Branscomb and James H. Keller, eds., *Investing in Innovation: Creating a Research and Innovation Policy That Works,* Cambridge, MA: MIT Press, 1998, pp. 85-111.

[27] Annalee Saxenian, *Regional Advantage: Culture and Competition in Silicon Valley and Route 128,* Cambridge, MA: Harvard University Press, 1994, p. 161.

> **Box C. Potential Risks and Guidance for Parks**
>
> A review of recent literature on the federal role in technology development reveals a list of potential risk factors that could jeopardize the success of the Ames Research Park. Among them:
>
> - An emphasis on developing a common vision may jeopardize the independence necessary to true innovation.
> - Excessive concern with intellectual property rights (unlikely to be of lasting value in the planned disciplines) may impede progress.
> - Cultural differences between researchers from federal laboratories, universities, and private firms must be addressed early and often to forge effective collaborations. For example, lab personnel are sometimes unfamiliar with the needs of commercial users.
> - The structure of collaborations must be planned with care. Helpful elements are sharing resources, sharing visions, and sharing physical space.
> - Special care must be taken in forging collaborations in areas distant from an institution's historic mission; e.g., educational outreach, entrepreneurial activity, joint degree programs.
> - Any CRADAs or similar agreements should be broad and flexible; specification of IP rights should not become an obstacle.
> - Special attention should be given to potentially competing goals—e.g., real estate returns vs. providing student housing; traditional vs. new missions; commercialization of results vs. free dissemination of knowledge (social returns).
>
> —Drescher, *Research Parks in the United States.* pp. 1-6.

Another advantage of Ames' location is that the presence of business expertise and facilities allow for the incubation of young businesses. Many parks today support their own incubators, which are designed to reduce business risk for researchers-turned-entrepreneurs by providing many functions: assisting young companies to use technology for economic development; moving discoveries from the lab to the marketplace; locating support services; and obtaining consultation, funding assistance, flexible leases, and office services.[28]

---

[28] Durso, "Home-Grown R&D," *op. cit.,* p. 2.

> **Box D. Regional Advantage in a Global Economy**
>
> "Paradoxically, regions offer an important source of competitive advantage even as production and markets become increasingly global. Geographic proximity promotes the repeated interaction and mutual trust needed to sustain collaboration and to speed the continual recombination of technology and skill."
>
> —Saxenian, *Regional Advantage*, p. 161

EDUCATION

Ames has also addressed in its strategic plan a pressing local and regional need—the education of tomorrow's science and engineering work force and the retraining of today's. Ames, with its educational and industrial partners, has planned an ambitious effort of outreach and on-site programs. As one scholar has written, "The greatest long-term threats to the Silicon Valley economy are . . . continued reductions in public funding for educational institutions—from its elementary and secondary schools to the sophisticated network of community colleges, state universities, and the University of California system—that jeopardize the rich supply of technical talent and the research base that have historically supported the regional economy."[29]

The editors of a leading treatise on technology have stated the challenge from a national perspective, describing our "inadequate technical and general education and inadequate retraining at all levels" as a significant obstacle to U.S. growth. "Needed," they write, "are education and retraining that can inspire a positive outlook toward science and technology, and an urge to maintain the American edge in technological competitiveness and entrepreneurial creativity."[30]

A REQUEST FOR ANALYSIS

To stimulate a full assessment of the issues relevant to this initiative, NASA's Administrator and the leadership of the Ames Research Center asked the Board on Science, Technology, and Economic Policy (STEP) at the National Research Council to convene a symposium to review the initiative.[31] Given the prominence

---

[29] Saxenian, *Regional Advantage, op. cit.,* p. ix.

[30] Nathan Rosenberg, Ralph Landau, and David C. Mowery, *Technology and the Wealth of Nations,* Stanford: Stanford University Press, 1992, p. 13.

[31] For additional background, see the Preface.

of Ames as a federal presence in the northern California region and the park's relevance to the Board's current review of U.S. government-industry partnerships, the STEP Board was pleased to respond positively to NASA's request. In particular, the Board sought to address the following issues:

- *Balancing Objectives:* How best to manage government-industry partnerships to accomplish NASA mission goals effectively and continue commercially relevant research, while at the same time properly managing—and balancing—access to federal facilities;
- *Developing Evaluation Metrics:* The importance of developing reasonable and accurate metrics to assess successes and failures in a complex, long-term undertaking such as a research park. Such metrics, of course, hold major interest for the managements of both Ames and its parent agency, NASA;
- *Ames' Interaction with the Private Sector:* The research park initiative as an element of Ames' interaction with the U.S. economy, especially small business and new start-ups, through its extensive supplier networks, existing partnerships, and expanding cooperation with industry.

THE REVIEW OBJECTIVE

These issues and the assessment challenges they entail are of great interest to policymakers and consequently to the National Academies' study of Government-Industry Partnerships. To consider the issues associated with the Ames initiatives in depth, the STEP Board organized a symposium to which it invited top members of the Ames and NASA management, leading academic experts on research parks, senior executives from the private sector, entrepreneurs in high-technology enterprises, experts from the world of private investment, and key congressional staff for an informal, but informed, dialogue.[32] This exchange enabled the Ames leadership to articulate its objectives for the park and benefit from the experience of the Committee through questions, comments, and occasional caution. It was recognized by all that the Ames S&T park is a work in progress and therefore one which could benefit immensely from an informed discussion of its objectives and likely challenges.

---

[32] A full list of participants is included as Annex C.

IV
PROCEEDINGS

Welcome

Henry McDonald
Ames Research Center

As director of Ames Research Center, Dr. McDonald welcomed the workshop attendants and offered a brief introduction the Center, which celebrated its 60th anniversary in December 1999. He showed an aerial view of the entire complex, which comprises roughly 2,000 acres.

Dr. McDonald observed that the overall mission of Ames is to support the objectives of its parent agency, the National Aeronautics and Space Administration (NASA). These objectives are classified under four major "enterprises": aerospace technology, human exploration and development of space, space science, and earth science. With the exception of earth science, the work at Ames is distributed fairly evenly across the major enterprises. Its work is supported by a budget anticipated to be approximately $600 million.

Ames also plays a major role is aviation operation systems, and is charged with developing somewhat more than 50 percent of the software that will be used to upgrade the national air transportation system to the so-called "free flight" mode. Other specific emphases include intelligent systems, high-performance computing, astrobiology ("in a word, the search for life") and information technology.

The human capital at Ames consists of roughly 3.5 thousand employees, of whom 1,500 are civil servants and 2,000 are resident contractors. Remarkably, of the total work force, 46 percent have advanced degrees, and nearly 60 percent of its scientists and engineers perform research. Graduate students and postdocs are resident during the summers. In short, Ames is a vibrant place to work, an important contributor to NASA's research enterprise, charged with one of the most stimulating research agendas facing mankind.

Opening Remarks

Zoe Lofgren
U.S. House of Representatives

Congresswoman Lofgren welcomed the conference participants, noting the great changes in Silicon Valley since her days growing up there in the 1950s. At that time, she said, there was "not much opportunity"; "the kind of innovation and success that is the hallmark of Silicon Valley simply did not exist." She noted that the area had benefited from the special combination of leading research universities, its innovative private sector, and federal research facilities.

She expressed her excitement about NASA's interest in developing a research park at Ames Research Center. In her view, it will provide a unique opportunity to develop and harness the talents of NASA, Silicon Valley firms, and the region's universities in a synergistic partnership that will benefit all parties. The NASA effort complements the University of California's plan to develop a regional education center in Silicon Valley. The University of California at Santa Cruz, she remarked, is leading the effort to build an education and research center that connects the resources and intellectual capital of the entire University of California with the specific interests and needs of Silicon Valley including NASA's research agenda at Ames Research Center. The University of California's plans include a new collaboration with San Jose State University and Foothill/ DeAnza Colleges that will specifically focus on bridging the digital divide in Silicon Valley and filling the workforce gap. In her view, this Center will also facilitate articulation and outreach activities with all the region's community colleges and create a distributed learning and research network that leverages technology and addresses the societal challenges of the Digital Age, including the changing demographics of California and the nation.

Congresswoman Lofgren, who serves on the Space Subcommittee of the

House Science Committee, said that Congress' support for science is sometimes insufficient, but that she sensed a growing understanding on both sides of the aisle of the importance of science funding. This funding can serve to educate our young people, to advance basic research, and to support efforts such as the Ames project in order to make sure there is an interface that works for the economy.

Panel I:
A Technology Vision for NASA

Moderator:
Edward Penhoet
University of California at Berkeley
and Chiron Corporation

NASA'S TECHNOLOGY STRATEGY

Sam Venneri

NASA[1]

Dr. Venneri began by describing the commercialization of technology as a major strategy for NASA, and praised the proposal to create a research park at Ames as a step in this direction. He then presented his vision of NASA's technology strategy for the future, involving "highly complex, first-of-a-kind missions which cannot be accomplished or afforded using current systems."

NASA's future mission challenges will require new systems for both space and Earth transportation. For the space shuttle, the goal for second- and third-generation vehicles is to increase safety and reliability. In the first-generation shuttle, some "3 million things" could go wrong. The chance that a problem will develop is about one in 250, which results in long intervals between missions and the need for some 20,000 people to prepare for each launch. (By contrast, a military pilot in combat has a one-in-10,000 chance of encountering a technical problem, and a commercial airliner a one-in-2 million chance.) NASA plans a third-generation shuttle that requires 50 or fewer people to process a payload and check out the system.

Similarly, in designs for exploratory spacecraft the agency is moving away from the single, large platform toward vehicles that are smaller and work in con-

---

[1] Dr. Venneri spoke via video connection from NASA headquarters in Washington, D.C.

stellation with each other, such as rovers for planetary surfaces and small craft to orbit planets.

These multiple, small spacecraft must be able to navigate for themselves, deal with uncertainty, and react to new conditions. This means that they must create information and knowledge from data, perform self-diagnosis and repair, and make decisions—in effect, to "think for themselves." Electronic circuits will repair and reconfigure themselves when necessary. A rover might be able to morph into some other state on a planetary surface; if the wheels get stuck, it may switch to another propulsion scheme to crawl or climb. It must be able to "live off the land" and utilize resources from the surface of asteroids or planets. The rover might "know" how to make propellants for itself and shelters for humans. This will require new ways for humans and machines to communicate and work together.

Such systems must be ultra-efficient, extremely durable, use little power onboard, generate power as needed, and move at low cost and high safety both around the earth and away from the earth. Systems must be highly distributed and comprised of interactive networks. Each system might consist of subsystems or units that can be damaged or broken apart and yet work together as constellations. In such a system, the failure of one unit does not mean the loss of the mission; failed units will be replaced or discarded.

The "Mission Triangle"

How does NASA plan to produce such hardy and versatile systems? Through integrating its three primary theme areas for the 21st century: biotechnology, nanotechnology, and information technology. Dr. Venneri described these three areas in terms of a "mission triangle," designed in collaboration with NASA administrator Dan Goldin. The integration of the three areas highlights and employs certain cardinal qualities of each:

- Biotechnology brings an ability to understand and simulate unique strengths of biological systems: e.g., an organism's ability to make exact copies of itself and to hybridize with other organisms. Thus the pieces of a constellated system might replicate themselves and/or hybridize with other pieces to continue functioning.
- The power of nanotechnology lies in its size—or lack of size. Nanotechnology means technology at the nanometer scale (literally, a billionth of a meter, at the scale of individual atoms). Engineering at the nanometer scale will reduce launch and power requirements and permit construction of smaller, cheaper, ultra-rugged systems.
- Information technology (IT) creates the means for communications, data storing and retrieval, and systems intelligence, effectively "coupling" the other two systems.

NASA's strategy—and Ames'—is to integrate the three systems to create evolvable, adaptable, self-repairing systems. This change, said Dr. Venneri, is as fundamental as moving from the vacuum tube era in the late 1940s to the transistor and semiconductor materials that have so radically transformed technology today.

The Nanoscale Approach

In engineering today, a typical example of a common manufacturing process is one that embeds graphic fibers in a polymer matrix to form fuel tanks and other objects. In a process at the nanoscale, this kind of process would move from the micron level upward; engineering process and failure methods are understood at that scale, and mechanistic fatigue and fracture are predicted at that scale. Instead of masking material and etching it away, as manufacturers do in conventional circuit-manufacture lithography, they would use "nanotweezers" and build up a material from atoms. Such materials actually have different physical structures and behaviors than today's materials.

Over the next 15- to 20-year period, said Dr. Venneri, nanoscale abilities will be developed and integrated into biological systems and the manufacture of engineering systems, connected by the "glue" of information technology. These relationships can produce a roadmap for a new national industrial base and its radical new products.

Techniques of nanotechnology (and nano-engineering) may accomplish truly revolutionary goals for NASA. They would begin with nano-structured sensors that have the ability to detect and characterize features at the quantum limit: single photons, cosmic particles, and molecules. These nanodevices and sensors would be designed to detect subtle signatures of life and to characterize deep space objects.

The next stage would bring ultrarugged nanoscale materials and structures that can withstand the harsh extremes of space. These would include microstructures for planetary and small body exploration, huge apertures to characterize extra-solar planets, and huge apertures to study phenomena under extreme conditions, such as black holes.

Finally, the third stage would feature a maturing of true nano-structural engineering, characterized by adaptivity and reconfigurability at the molecular level and merged software and hardware for biomimetic systems that are responsive to changes in both internal and external conditions. These advanced nano-systems would allow the development of self-repairing spacecraft, self-configuring space systems to optimize mission returns, biomimetic systems for robotic exploration, and space system lifetimes of decades to centuries for interstellar exploration.

Nanobiotechnology

In nanobiotechnology, the nanoscale approach is enriched by applying the capabilities of biology. Each research effort would advance nanostructural engineering to produce high strength/mass ratios. Typical efforts would be to emulate the structure of natural structures, such as spider silk, and to produce natural or artificial biomimetics.

Other joint capabilities might include nanodevices and sensors that go beyond binary, silicon-era computers and into the era of quantum, DNA, or protein-based systems that operate at different scales. They might also include parallel processing that starts to mimic how the brain processes information. Devices at the atomic level may even be able to monitor body systems at the cellular level. These devices could work in clusters and communicate with each other as they observe, for example, cellular damage and mutations. NASA is presently working with NIH on a nanotechnology to monitor signs of early ovarian cancer, which is almost impossible to detect with conventional technology.

Information Technology, Nanotechnology, and Biotechnology

NASA will make use of any appropriate IT systems developed by the commercial IT industry. Because many space systems have little commercial application, however, the Agency anticipates the need to develop many IT systems of its own. In particular, it plans to explore four specific areas that couple nanotechnology with biotechnology:

- The first is fundamental research in automated reasoning—the ability to embed intelligence in systems through software techniques (or "soft" computing). These are systems that reliably make and execute decisions that traditionally require human intervention. Constellated systems will require neural net technology, genetic algorithms, and fuzzy logic, and a substantial move away from the hard, deterministic numerical computing of today.
- The second area is "human-centered computing," systems by which humans deal with machines in ways that amplify what either can do. This may be thought of as a matrix: an intelligent agency computing with humans, perhaps even in a natural language. Such a matrix would allow humans to work with all their senses, not just a Windows-type environment.
- The third area is intelligent data understanding. This involves autonomous techniques to transform data into information, information into knowledge, and knowledge into understanding. A growing problem for NASA is the overwhelming quantity of data that is produced by systems operation, remote sensing, and other processes. This information must be presented in ways the human brain can process; in other words, as a knowledge base. This "data fusion," or data product development, has the goal of maximizing human interaction with knowledge.

- The fourth area is revolutionary computing that moves beyond the silicon era to provide a platform for the development of future "intelligent systems." Next-generation computing systems may be quantum-based, biology-based, photonic, or (very likely) a hybrid of these systems. The computers of the future may be the size of the human brain and function at power levels of watts, not kilowatts—much like biological systems.

The computing environment of the future might be very different from the familiar present. This environment might be three-dimensional, and it might allow people to use all their senses. It might also mean that the "person" we see and communicate with would not be human at all, but an intelligent agent manifesting in a cave vision dome environment. Geographically dispersed teams would come and go from this virtual world, where intelligent agents would interact with each other and with humans to develop complex products or knowledge.

A practical exercise for this kind of computing is to develop a new idea through virtual means—to move a conceptual, detailed engineering design through manufacture, use and its entire life cycle—in one year rather than the five years required today. Every step would be designed and rehearsed in virtual space before the first piece of hardware is cut.

NASA hopes within the next 20 years to be able to extend this ability to the nanoscale, mimicking and manipulating atoms and molecular biological structures at the atomic scale by virtual means. Artificial DNA and its components, for example, would be part of such a design space, starting with the fundamental building blocks of nature.

Self-Healing Structures

For physical structures, NASA's goals include self-healing organic binders for structural composites, ionomers that can heal cracks with ultrasonic or microwave energy, and the capability to regrow materials and repair damage in load-carrying structures. One biological model is living bone, which is able to regenerate and repair itself by adding material around stress concentrations before cracks grow too far. This might be mimicked by adding tubes of material adjacent to load-carrying fibers so that the material would be available to ooze into new cracks and repair them by hardening in place. Such techniques would be applied to aircraft and spacecraft to keep structures airtight and prevent failures.

Dr. Venneri said that a typical objective in biomimetic engineering might be to emulate the efficient skeletal structure of a frigate bird, whose wings span some seven feet but whose skeleton weighs only four ounces. The secret is the use of a hollow, tubular bone structure. To use biomimetics for such "novel" structural designs, one would design hollow tubes that resembled wing bones, rather than using the traditional spars and ribs of aircraft wings. To complete a strong but light aircraft structure, engineers would assemble a thin skin from optimized lightweight material and design an aerodynamically efficient, thin airfoil.

Dr. Venneri concluded by saying that Ames and the surrounding region is an appropriate base from which to pursue this technological vision and its three main emphases. "This synergistic coupling offers a revolution that any of these areas on their own would not begin to achieve," he said. "We have the potential for self-assembling electronics, for artificial DNA, for a third-generation launch system that is truly a thinking vehicle. With a distributed nervous system it can self-certify, it can talk to people, it can warn of a structural part going bad and ask permission to replace it. Through partnerships in the universities and private firms in this region, we can become the wellspring for this new technological level. Ames is our seed gene for really bringing this together in the agency."

AMES' TECHNOLOGY STRATEGY

Henry McDonald
Ames Research Center

Dr. McDonald extended the discussion of NASA's three primary theme areas, explaining that all the elements of the agency's new scientific and technological direction have significant leadership and representation at Ames. The origins of this leadership have much to do with Ames' location in Silicon Valley. The center began early to build up a strong infrastructure and staff in information technology, and is now the lead NASA center for IT. It also became a national leader in advanced computing, and extended its work to artificial intelligence. Dr. McDonald emphasized the world-class science being done at Ames, where researchers have won two Feynman Prizes, published more than 100 scientific papers, and earned four patents since 1996.

When Ames was asked by NASA Administrator Goldin to revise its strategic plan, it was logical for Ames to continue its focus on IT. In recent years, the center had begun to build up expertise in life and microgravity sciences, adding significant strength in biology. It is NASA's lead center for astrobiology. Expertise in the third theme area, nanotechnology, grew out of Ames' supercomputing mandate.

Dr. McDonald expanded on several points made by Dr. Venneri, including the use of nanotechnology to produce very light launch vehicles. Over the last 25 years spacecraft have become lighter by roughly an order of magnitude (from 1000 to 100 kg) and in the future will shed another order of magnitude (to approximately 10 kg). Rather than more 1000-kg Cassinis, whose design might require 15 years and whose failure would mean frustrating loss for its designers and the agency, NASA will build a larger number of smaller, less-expensive, more-reliable vehicles.

He also discussed a change in the way the space agency quantifies its missions. For the past few decades it has focused on launch mass and reliability. A new approach is to look at capability. For a communications satellite, capability might be defined as throughput in bits per second times the number of channels on board times the expected lifetime of the satellite. To make dramatic breakthroughs, satellites will require more capability of the kind that information technology and nanotechnology are likely to provide.

A 20-Year Vision

Dr. McDonald reiterated Ames' 20-year vision in terms of "self-assembly, self-diagnosis, self-healing—all processes that the biological world knows how to do very well." He enumerated the biological processes of interest to nanobiotechnology as follows: replication, production of useful output, growth, evolution, and repair (including reconstruction, reconfiguration, and replacement).

Ames is looking at three elements to understand these processes. The first is the carbon nanotube junction. This has been studied theoretically, using computational electronics, to see if such a junction is possible and how it might perform electronically. The second element is self-assembly at the molecular level, with potential electronic uses. Ames biologists are studying a donut-shaped molecule found in a certain "extremophile," an organism that flourishes in extreme conditions (in this case hot water). The biologists found that by trimming the donut they could get it to self-assemble into an array. They are now trying to put into the center of each of these molecules a metal atom or some other atom, perhaps one that can accept a photon; it might then be possible to have an array that could read and write at the molecular level. The storage implications of such a capability would be enormous for some missions. This work brought to NASA's attention the convergences of biology and how it affects information technology.

Dr. McDonald offered another example of the potential power of harnessing biological function in the service of technology. If one wants to build a device with a million transistors for use in an extreme environment, such as heavy radiation, one can be certain that some of these transistors will be damaged. They can be protected by shielding, but shielding is heavy. What is needed is a degree of self-healing—a process developed in biological organisms.

Autonomy and Intelligent Agents

Space systems also need (like organisms) a degree of autonomy. A major issue is the total number of decisions that have to be made on a mission, and how often they have to be made by a human. For a sensor on or near Mars, communication with humans on Earth can take as long as 30 minutes, so a successful sensor is one that goes for long periods without requiring human input or response. Similarly, the robot colonies planned by NASA need to be able to make decisions when they encounter a problem. The issue of autonomy is related to human-centered computing, a combination of humans and computers in which computers ("intelligent agents") take increasing degrees of the responsibility. An intelligent agent is one that reliably makes decisions ("autonomous reasoning") with limited human intervention.

NASA is projecting that the proportion of autonomous decision to human decisions will increase over the next two decades, along with mission complexity (i.e., the number of decisions to be made). A rough progression might include the following mileposts: 1) Predictive diagnosis, for planetary exploration (2003); continuous response to unknown environments, e.g., for the Europa Submarine (2015); collaborative intelligent agents, for robot colonies (2016); and science-driven operations for fleets of spacecraft (c. 2020).

The concept of intelligent agents is also being developed at Ames for the commercial air traffic control system, which must be prepared for a doubling of air traffic in the near future. The number of controllers cannot be increased much further because it is limited by a system that requires inter-controller communication. A revised system is needed that depends less on person-to-person communication and more on autonomous, computer-aided decisions by pilots.

Finally, Ames is challenged to extend its computing expertise into "intelligent data understanding." The agency has more data than it can access efficiently. For example, the space shuttle has a 25-year history, and enormous amounts of information are spread over a half dozen distributed sites that can not be accessed simultaneously with available browser technology.

In addition, the agency will be downloading terabytes of data each day from space satellites imaging the Earth, progressing from the current Landsat to the more advanced Terra satellite. One goal to identify the features of the Earth in detail, but the size of data sets will rise from about 10^6 presently to 10^{15}. "Responsiveness"—the number of queries that can be completed in a day—will have to increase from only about one at present to a million for the Remote Lunar Vehicle, scheduled for operation from 2010 to 2015. Other systems, such as the Space Interferometer (2005-7) and the Earth Sensor Web (2020-30), will make additional demands on data handling and understanding.

The computing problem is compounded by the need to put high-capacity computers into space. Conventional computers must be protected, which adds a lot of weight, and they are hitting their theoretical performance wall. The agency urgently needs to find a more effective form of computing to do the science and exploration it wants to do. In particular, the computers of the future must be extremely robust and able to fix or work around the glitches that characterized the early decades of space exploration.

In summary, Dr. McDonald concluded that Ames has appropriate expertise to strengthen its leadership role in research. The challenges involved in moving NASA to the required higher level of technology are clearly beyond the resources of any single center. Ames proposes to do so by utilizing its favorable physical location and creating strategic partnerships with government, industry, and academia in areas of high mission priority.

QUESTIONS & COMMENTS

Budgetary Commitment?

In response to a question about budget, Dr. McDonald said that a substantial amount of money will be dedicated toward establishing research partnerships at Ames. For example, approximately 25 percent of the budget of the new intelligent systems program will be available for extramural activities with universities and a further similar amount for academic-industrial-government initiatives. The same is true for the fundamental biology initiative.

The Imperatives of Physical Location

Dr. Wilson asked how important was the physical location of Ames. Dr. McDonald said that the presence of a single site is critical to generate the interactions that occur between investigators at various levels, from graduate students to principal investigators. This is greatly strengthened by personal acquaintance and interaction. Relationships, once established, can then be sustained by telecommunications. Alternatively, interaction can be initiated remotely and then strengthened by physical proximity. The astrobiology institute, for example, is now a virtual institute which will be expanded with on-site facilities so that various participants can come for varying periods and graduate students can do doctoral dissertations.

Ames' Competitive Advantage

Dr. Penhoet pointed out that NASA's mission triangle is now a common theme in universities, including Stanford and MIT, and asked whether Ames had a competitive advantage.

Dr. McDonald mentioned several: 1) Ames is one of the few NASA centers employing significant numbers of people on-site in all three areas: 40-50 people full time in nanotechnology, 150 in the biological sciences, and 150-200 in information technology; 2) Ames' mission requirement to solve these problems provides focus and drive; and 3) because NASA pays its corps of researchers by annual salary, they do not have to spend time writing grant proposals, a significant advantage. Dr. McDonald was also asked whether Ames would be primarily a developer of technology or an aggregator of technology developed elsewhere. He said that Ames will seek out technologies already developed by industry and academia. However, much of its research would focus on areas that are necessary to NASA missions (e.g., space-based computing) but do not attract other groups. Also, because of its range of personnel, Ames can take a technology all the way from the conceptual stage to development and application.

Panel II:
Research Parks:
Concept, History, and Metrics

Moderator:
David Audretsch
Indiana University

PRESENTER

Michael Luger
University of North Carolina at Chapel Hill

Dr. Luger, who has studied and written extensively about research parks, addressed three topics: 1) background and context, 2) design issues, and 3) ways to measure success.

The number of research parks world-wide has grown dramatically since the 1950s, when the pioneering Stanford Industrial Park was established in Palo Alto, California. Depending on definition, there are hundreds or even thousands of parks in more than 60 countries. There are 295 members of the Association of University-Related Research Parks, several hundred members of the International Association of Science Parks, and dozens of members of regional science park organizations.

The International Association of Science Parks defines a research park as one that has operational links to research centers, universities, and other institutions of higher education; is designed to encourage the formation and growth of knowledge-based industries or value-added tertiary firms; and has a management team actively engaged in transferring technology and business skills to tenant organizations. The most successful parks can have a profound impact on a region and its competitiveness. For example, Research Triangle Park (RTP) in North Carolina is credited with generating (directly or indirectly) some 25 percent of all jobs in the region, and altering the region's basic economy.

Varieties of Parks

Within this broad definition are many variations of the research park model that differ by structure and function:

- *Research parks*, such as Research Triangle Park and Stanford Industrial Park, focus primarily on R&D and may exclude manufacturing or assembly.
- *Science and technology parks*, such as the Centennial Campus in North Carolina and the University of Utah Research Park, focus on translating the results of research into new products or processes for commercial applications.
- In developing countries, many high-tech "industrial or agricultural parks" are essentially *groups of firms* that assemble and produce high-tech products.
- *Warehouse and distribution parks* are basically "big boxes" that concentrate on warehousing and distribution and increasingly develop advanced information technology and logistics.
- *Global transparks*, with examples in Kinston, North Carolina, and Thailand, operate just-in-time production facilities near decommissioned airfields to move goods around the world very quickly.
- *Headquarters parks* focus on sales functions and administrative activity rather than R&D.
- *Eco-industrial parks* tend to be regional affiliations of firms linked in order to use each other's inputs and byproducts in ways that reduce environmental impacts; a prominent example is found in Kalundborg, Denmark, 75 miles east of Copenhagen. This park began spontaneously when members sought ways to reduce costs and meet regulatory requirements.

Several other kinds of research-oriented facilities share similarities with research parks:

- *Research and technology centers* are physical facilities that may or may not be located in parks; many parks are anchored by such centers. The NSTDA park in Thailand is built around four R&T centers. RTP in North Carolina is built around biotechnology and microelectronic centers. A park in Palestine is focused around centers of software and hardware networking support.
- The *technopolis* (or science city) is a larger region that is developed around several high-tech elements, including but not limited to research parks and R&D centers. Prominent examples are Tsukuba in Japan and Taedok Science Town in Korea. In southwestern China a technopolis is planned in the metropolitan city of Chongqing, along the Yangtse River, including

Chongqing University and some 25 other institutions of higher education or research centers in the region. Dr. Luger suggested that Ames could become the high-tech center for a technopolis in the northern California region.

Trends in Research Parks

Dr. Luger went on to describe four recent trends among research parks during the last several decades.

Concentration on Key Sectors

Parks have concentrated on one or more key technology sectors, for the following purposes: a) to provide greater focus, b) to strengthen marketing ability, and c) to make a greater contribution to science. In the case of RTP, biotechnology, telecommunications, pharmaceuticals, and software have become the major foci. In Thailand, the foci are biotech, materials sciences, electronics, and informatics; in Palestine, software development and network support. Dr. Luger said that the list of complex technological objectives that had been discussed by Ames leadership might be too ambitious (including carbon nanotube junctions, self-assembling molecular structures, self-healing transistor arrays, intelligent agents, intelligent data understanding, and extremely robust computers), but that Ames certainly hold competitive advantages in some of these areas.

Linking to Clusters

The intensified competition for R&D capacity has prompted more parks to tie into existing and emerging industrial *clusters*. These are groups of firms and related institutions whose competitiveness depends on the competitiveness of other members of the cluster. Clusters may form among businesses related through input-output linkages (such as the automobile cluster around Detroit) and among firms that share the same labor needs, skill sets, or output markets. In forming a cluster, research parks must be supplemented by cost incentives and training programs.

The standard model for clusters around research parks has four features:

a) R&D activities appropriate to nearby industries;
b) technical training to produce skilled technicians;
c) enabling services (such as network brokers, regulatory assistance, entrepreneurship, and technology transfer) that make the system work more smoothly; and
d) continual modernization and upgrading of technology. The technopolis growing in Chongqing, China, is a cluster in the sense that it adds R&D capacity around a motorcycle production facility.

In North Carolina, the RTP just finished its 2030 visioning process, which describes large and growing clusters in communications, software, motor vehicle manufacture, and information technology. Planners must now ask what support services and R&D facilities are needed to allow those clusters to continue to gain international competitive advantage.

The "Green Door" Concept

A third trend in research parks is the extension of the university-industry connection into something called the "green door" concept: scientists from universities, who are busy teaching and advising grad students, are offered convenient access to industry-supported labs that focus on commercializing the results of research. A leading example has been initiated at North Carolina State University in Raleigh. The green door area is called the Centennial Campus, developed on 500 hectares of land next to the university. The internationally known school of textiles moved there, and the engineering department built a graduate research center as well as incubator space for private businesses (called partners). Incubator space is reserved for companies that can demonstrate meaningful relationships with students and faculty doing research. In just a few years the success of this venture has far surpassed expectations, linking more than 900 scientists and engineers with 60 partners and producing a number of patents and licensing agreements.

Virtual Parks or "Collaboratories"

The fourth trend is the development of virtual parks, sometimes called the NSF "collaboratories." These virtual parks link scientists and engineering researchers around the world in real time via information technology. The intention is not to bypass person-to-person contact, which will always be a necessary foundation for any collaboration, but to supplement it. This strategy is proving increasingly productive in sustaining research partnerships.

Parks differ not only with respect to strategies and structures, but also by the incentives they use, the services they offer, and their ownership and leasing practices. Each park is unique, and must choose the mix of elements that make sense in the local context. Local traditions and legal constraints play large roles in the success of any park.

In general, however, virtually all research parks try to accomplish the same broad set of objectives: incubation, training, services, and research. Successful parks are those that are able to blend their strategies for pursuing all four of these objectives.

Questions of Park Design: Conditions for Success

Park organizers have sought to define and measure success in various ways. Two questions are always important: first, *ex ante*, how can one know whether a park should be built, and second, how does one know whether a park, once built, is successful. These fundamental policy questions have been studied for several decades.

Answering the first question may begin with a market analysis and consideration of the following issues:

- Is there sufficient demand to make the model work?
- Is a park the best way to achieve a set of objectives?
- What are the primary objectives for a planned park?
- What resources are available? (Many parks have foundered for lack of adequate support.)
- What's the expected payback period? (Parks may develop slowly.)
- How many public benefits are expected? (This question is especially important when public funding, such as transportation funds, are involved.)
- Does the proposed site have the right fundamentals to make the location attractive to industries? (Ames, for example, offers pre-existing technological prominence, extensive physical facilities, and a strategic location next to Silicon Valley.)
- Is low-priced land available? (The most successful parks, notably Stanford and RTP, had available land—as does Ames.)
- Is there access to customers, to a labor supply with appropriate skills, to physical infrastructure (road, sewers, electricity, gas, etc.), a knowledge infrastructure (e.g., universities), capital, and a good quality of life?
- Is there experienced and visionary leadership?
- Is there political and citizen support? (Many parks that failed lacked long-term commitment of political, business, and government leaders.)
- Are governmental and other organizations prepared to take supportive policy actions and be counted as stakeholders?
- Is the concept based on a realistic reading of the economy? (There are limits to the accuracy of economic prediction, but a generally accurate assessment is important.)

Measuring the Success of Research Parks

The second question, how to measure the success of a park, cannot be answered by any single method, but a combination of several methods usually brings useful information.

Expectations vs. Outcomes

One way to look at success is through "effectiveness analysis": measuring outcomes against expectations in the near and long terms. Near-term goals tend to be easily quantitative: numbers of new tenants, employees, and patents. In the long term, goals are often more lofty and sometimes less easy to measure: creating new industries, not just attracting more of what exists; creating new wealth, not just new jobs; developing a critical economic mass, not just arresting decline.

Attribution Issues

Attempts to evaluate the success of a park are hampered by problems of attribution: to what extent are effects attributable to the development of the park and its programs? Would these effects have occurred anyway? How should costs be allocated; is technical training on or near the park a cost to the park? Once a development pattern is set, evaluation is not a very useful tool for changing course, because the development pattern is already set. Experience to date suggests that adaptability and flexibility during development are important determinants of whether a part succeeds.

Points of Comparison

Several quasi-experimental methods can also be helpful, if not precise. One is to compare a park with another place that is similar in every way except for a park. A second method is the use of case studies which, although they don't give objective data, can offer rich detail. A third method is econometric analysis. With hundreds of observations, one can do a standard regression and try to attribute outcomes to various activities. In addition to the imprecise nature of these methods, another challenge is how to allocate costs. Is an investment in technical training, for example, a cost to the park or to the public sector?

The Need for Accountability

Despite such uncertainties, Dr. Luger pointed out, accountability is required—by GPRA, by NASA's budget, by Congress, and by the court of public opinion. Benchmarking has become a routine exercise in science and technology policy, and research parks are no exception. He closed by offering a sobering statistic: one-half of the parks that were initiated between 1960 and 1990 have ceased to exist. Of the remainder, half had to shift their emphasis away from research in order to survive.

In terms of efficiency, or cost benefit analysis, few parks generate tangible benefits that exceed their costs, for a number of reasons. Some parks are judged too quickly; long-term goals may take 20-50 years to come to fruition. In the case of RTP, the park and affiliated universities were successful in transforming a rural, low-skilled, textile- and tobacco-based economy into a high-tech magnet for industries—but the process took 40 years.

Reasons for failure include increasing competition, an unrealistic concept, the lack of certain key ingredients, too narrow a definition of benefits, and too broad a definition of costs. Conversely, success factors for successful parks include leadership, vision, "deep pockets," patience, good timing, good luck, appropriate services, and meaningful connections with universities. Proximity to universities is not sufficient; parks must strengthen universities as well as draw on them, so that the competitiveness of both is increased.

QUESTIONS & COMMENTS

Pursuing this point, Chancellor Greenwood asked what kinds of university connections are most likely to advance the goals of research parks. Dr. Luger said that the best relationship is a "two-way street." The university benefits because high-tech companies tend to contribute funds for research, make scientists and engineers available as adjunct professors, build facilities on campus, and employ graduate students. The park benefits in drawing on the university's intellectual resources, gaining access to and sometimes commercializing university research, and forming joint research projects. In addition, with respect to virtual parks, industry researchers gain from association with university researchers, who tend to be better connected globally with others in the field.

DISCUSSANT

Susan Hackwood

California Council on Science and Technology;
University of California at Riverside

Dr. Hackwood, who has held positions in academia, private industry, and government, brought a broad perspective to the discussion. She commented that the Ames plans constituted "an extremely interesting venture," and found the research areas under consideration to be "at the leading edge of what's happening in science and technology." She said that the challenge to Ames is not so much to generate outstanding research as to find a niche that is not already occupied in this high-tech valley.

The Research Wealth of California

From her position on the California Council of Science and Technology, she perceived California to be virtually "a country in itself" in terms of size, wealth of research activities, and contribution to GNP. The Council was created by the state about 10 years ago to represent and support the state's research enterprise. The organization is comprised of 120 people: a board, a council, and a number of fellows, half of whom are members of the National Academy of Sciences and six of whom are Nobel Laureates. The Council's function at the state level is similar to that of the National Research Council at the federal level—to represent the state's interests to the state government, to the federal government, to the Congressional delegation, and to the academic institutions and industries that comprise the council.

The CREST Report: Higher Research Per Capita

The Council has completed a major study on science and technology in California which examined the following components: overall S&T effort, high-tech industry, academic research, state S&T policies, federal laboratories, foundation support, venture capital, and human resources. Then it asked whether these components were sufficient to sustain the state's research momentum and economic growth in the future. The result of the study, the California Report on the Environment for Science and Technology (CREST), was published in November 1999. It revealed that California, with 12.5 percent of the nation's population, conducts about 25 percent of the nation's R&D in science and technology.

Two Disturbing Trends

However, two disturbing conclusions emerged as well, both of which suggest a need for substantial changes and carry important implications for the growth of the Ames research park.

Falling Federal Funding

First, the report revealed a significant change in the kinds of research being supported in California. At first glance, the situation seems healthy; total funding for R&D has risen steadily (except for a pause during the 1990-91 recession) since 1977. This upward trend is a function of a large increase in R&D funding by industry, which is essential to sustain innovation and economic productivity. One result of this upward trend is that the number of patents granted to California researchers in high-technology fields rose considerably faster between 1980 and 1996 than the number of patents in other high-technology states such as Massachusetts, Michigan, and New York. For example, the number of patents issued for California inventions doubled in electronics and tripled in biotech.

While industry funding for R&D has risen rapidly, however, federal funding has "nose dived." Because the federal government is the primary source of support for basic, long-range research, especially in university laboratories, these forms of research have suffered disproportionately. Industry, by contrast, tends to fund research that is shorter-term and product-oriented.

Lower Funding for S&T Disciplines

The second disturbing finding of CREST is that most California academic institutions, with the exception of those in the top rank (Caltech, Berkeley, Stanford), have suffered a significant erosion of funding for the S&T disciplines, especially engineering. Overall, the number of engineering graduates from California institutions has decreased 9 percent over the last 10 years. Among state university campuses, which produce the of bulk of the S&T workforce, the number has dropped 25 percent.

The state depends on these graduates to fill the high-tech jobs of the future, said Dr. Hackwood, such as those contemplated at Ames. California is already an importer of engineers from other states, but even higher numbers will be required to sustain the state's high-tech growth. This educational deficit forms a serious challenge not only for universities, but also for community colleges, secondary schools, and, significantly, the K-12 schools that produce the workforce of the future.

In conclusion, she said, the "miracle is not guaranteed." Reasons include the following: Other states show increasing and stronger commitment to support of R&D; California ranks 32nd in R&D state government funding per capita; and California is not preparing its future citizens for high-technology jobs.

QUESTIONS & COMMENTS

In response to a question from Elizabeth Downing about the reasons for the decline in technology graduates, Dr. Hackwood described the drop as part of a national trend. California is especially hard hit because of its heavy reliance on research. The reasons for the decline, she said, are many, and begin with the decreasing amount of money available to support graduate students. At the baccalaureate level, schools face rising costs for engineering programs, rapid obsolescence of equipment, and the loss of faculty to private firms. At the associate level, the total number of students entering the system is rising, but fewer choose science or engineering (with the exception of some areas of health science). The K-12 grades suffer from inadequate overall funding—especially for science and mathematics—and a demographic shift toward pupils from cultures who do not traditionally enter S&T.

Dr. Wessner of STEP noted that a reluctance to spend public funds in strengthening education for the current generation almost certainly jeopardizes the prosperity of the next generation, and asked if the Council attempts to raise public awareness of this risk. Dr. Hackwood said that the Council is now beginning a concerted effort to influence state policy on this point.

Dr. Wessner also noted that a STEP report had identified specific problems in K-12 education, and he asked about ongoing efforts to change the system in California.[2] Dr. Hackwood observed that any changes to the system encounter political opposition and a reluctance to spend money. She pointed out that the United Kingdom had improved its K-12 system "at the cost of very significant changes in the governance structure and the equivalent of 'changing the constitution' for K-12." She said that the state is now looking at how to change policy in California, to change institutional funding, to increase enrollment and retention, and to be sure that pupils in K-12 "have science and engineering on their event horizon."

Dr. Greenwood added that the K-12 issues raised by STEP's report, among others, are beginning to be actively addressed in California, both by the governor and by a new, $300 million program for outreach from the UC system. Current efforts include identification of underperforming schools, means of raising performance, and ways to address the fact that the new workforce is comprised largely of people who have not traditionally entered S&T fields.

Dr. Behrens, of Robertson Stephens Investment Management, asked about the source of the pool of technology labor. Dr. Hackwood said that the pool is made up of two groups. The first includes those responsible for innovation, who constitute a small minority. The deficit occurs in the second group, which includes technicians and others at the associate and baccalaureate levels who make up the bulk of the workforce. These must be imported from other states, since California does 25 percent of the nation's research but only produces 9 percent of its engineers. She added that foreign S&Es, including those on H1B visas, are relatively few in number, and don't relieve the state of the need to produce its own workforce.

Mark Weiss of Xerox, who has served as a director of several start-up companies in Silicon Valley, reemphasized the significant deterrents to moving to the area, including the high cost of living, the shortage of affordable housing, and the overburdened transportation system. Dr. Luger responded that these same concerns were described when he was evaluating research parks in the area as long as 15 years ago, when companies feared they would not be able to import enough technicians to support company growth. He said that on-site housing at Ames could at least help alleviate the problem at this site.

---

[2] See National Research Council, *Improving America's Schools: The Role of Incentives.* Eric A. Hanushek and Dale W. Jorgenson, eds., Washington, D.C.: National Academy Press, 1996.

Panel III:
The Ames Research Park:
Goals and Metrics

Moderator:
Patrick Windham
Stanford University and Windham Consulting

THE AMES STRATEGIC PLAN

William Berry

Ames Research Center

Following Dr. McDonald's description of Ames' research objectives, Mr. Berry discussed a series of proposals for research partnerships, education, outreach, regional issues, and site construction. In general, Ames administrators intend to phase out some operational functions in favor of strengthening and extending research capabilities. The broad strategic objective is to "develop a world-class shared-use research and development campus in association with government entities, academia, industry, and non-profits."

In particular, the plan has two components. The first is to support NASA's overall mission in three areas: 1) advance NASA's research leadership; 2) enhance the Agency's education, outreach, and advocacy efforts; and 3) create a unique community of researchers, students, and educators. The second component is regional involvement. Ames would invite regional participation in planning and in key projects or partnerships, and organize its activities in ways that are consistent with Bay Area interests, including environmental, transportation, educational, and economic interests.

The strategic plan proposes extensive renovation and expansion of Ames' extensive physical plant, which is located at the heart of Silicon Valley near the cities of Mountain View and Sunnyvale, between the wetlands of San Francisco Bay and Route 101. The Ames property is adjacent to the Lockheed Martin Missile and Space Co. and close to numerous high-tech companies and research facilities, including new components of Microsoft and Netscape.

As a result of the Base Realignment and Closure Act of 1988, the original NASA/Ames tract of 500 acres was increased by 1,500 acres in 1994 by the addition of Moffett Naval Air Station. Ames is now the host agency for several other governmental organizations, including the California Air National Guard, Army reserve units, small active Army units, and a Federal Emergency Management Agency unit, which operate on a shared cost-pool basis. The airfield easily accommodates modern aircraft (e.g., Boeing 747s).

Objectives and Strategy

Mr. Berry began to develop the current strategic plan in January 1997. He emphasized that the objectives of the plan support the NASA mission. The plan's primary objective is to extend and deepen the research and development capabilities of Ames through R&D partnerships with industry, universities, and other entities. These partnerships would primarily be in the areas Mr. Venneri described earlier, that is, information technology, nanotechnology, and biotechnology, with complementary programs in astrobiology.

Related objectives are to:

- create new and unique research facilities and other related physical facilities that further the mission of NASA;
- conduct education and outreach programs, in partnership with universities and school systems, for multiple purposes—
  - enhancing the education of the Ames work force;
  - creating graduate, postdoctoral, intern, sabbatical, and other opportunities for visiting scholars;
  - developing public educational programs on site; and
  - strengthening science and technology in regional school systems;
- develop the site in ways that are consistent with regional goals and that promote employment, sound land management, good environmental practices, clean and efficient transportation, and economic development;
- make available some of Ames' former military housing for researchers; and
- manage in appropriate ways the portions of the site that have been designated as historic districts and buildings.

The Ames Approach to Partnerships

Mr. Berry emphasized that the Ames partnership strategy is based on its legislative mandate, the Space Act of 1958. In delineating the authority of NASA, the Act permits the agency to enter into "contracts, leases, cooperative agreements, or other transactions as may be necessary in the conduct of its work and on such terms as it may deem appropriate...with any person, form, association, corporation, or educational institution." According to consultations with NASA's counsel general, the center requires no additional authority for its planned activities. The Act also authorizes NASA to recover costs it may incur in supporting a collaborative activity, such as providing fire and security services to its partners.

The partnerships currently envisaged include the following:

- The formation of any partnership must be an open and fair process that entertains all potential offers, with metrics that show true accountability.
- Partners can create (or renovate) their own new facilities on the Ames property, including subleasing to tenants approved by NASA.
- In return for providing land and a unique relationship with NASA researchers, Ames would require some of its partners to reinvest funds gained through partnerships in future collaborative research activities at Ames. NASA will cost-share its own activities when it is directly part of a collaborative activity. Thus the partnerships would be fueled primarily by nonappropriated funds.

Ames has entered into memoranda of understanding with potential partners through 2000: the University of California system, led by UC Santa Cruz; Carnegie Mellon University; San Jose State University; Foothill-DeAnza Community College District; the National Association for Equal Opportunity in Higher Education (NAFEO); and Lockheed Martin. The center is presently "moving down parallel paths" with each potential partner toward a "convergent point" that satisfies both NASA's primary objectives and the objectives of each partner. Mr. Berry described the major features (includes post-briefing calendar year 2000 updates) of each partnership as follows:

- *University of California at Santa Cruz (UCSC)*: This would be the lead UC campus, which NASA views as the portal into the UC system. Designating a lead campus in the Research Park is intended to reduce competition for students and provide a more integrated approach to educational projects. The UC System has designated Ames as its preferred site for its Silicon Valley Center, a UCSC-managed research and education campus. Ames anticipates benefits from joint research projects of mutual interest and relationships with graduate students and postdocs. "UC and NASA scientists will work together on advances in science and technology that will drive new industries and provide new products benefiting California's economy," said UC President Richard Atkinson at an October 25, 2000 press conference announcing the partnership. "UC Santa Cruz will serve as a portal to the UC system for Silicon Valley to connect UC's intellectual resources with the specific interests and needs of Silicon Valley, NASA, the state and the nation," Atkinson said. UCSC and NASA share many research areas of interest and strengths, such as biotechnology, nanotechnology, planetary sciences, and astrobiology, and other UC campuses have expertise in information technology. In addition, research is expected to be conducted on issues of social justice, education, labor, and economics, among other topics.

- *UCSC, San José State University, and the Foothill-De Anza Community College District*: These schools have formed the "Collaborative," an unprecedented academic partnership to address Silicon Valley's critical education and workforce needs through joint research and education programs to be located at the NASA Research Park. "By crossing traditional boundaries, our collaboration with San José State and Foothill-De Anza will leverage our collective strength, provide innovative programs and services and produce results of value to Silicon Valley and the State of California," said UCSC Chancellor M.R.C. Greenwood at the September 5, 2000 press conference announcing the partnership. "The NASA Research Park will provide the optimal environment for collaboration. Through these collaborative programs, we look forward to building world-class facilities that will house teaching, research, and economic development programs for the Silicon Valley and the State of California," said San Jose State University President Robert Caret. The Collaborative has already received a $100k grant from the Packard Foundation to begin planning the Teacher Institute component for the proposed California Air and Space Center (CASC), an independent nonprofit, that will renovate Historic Hangar 1 into a world-class science and technology learning center. The CASC will partner with NASA and the Collaborative in a number of areas.
- *National Association for Equal Opportunity in Higher Education (NAFEO):* NAFEO is the national umbrella and public policy advocacy organization for 118 of the nation's historically and predominantly Black colleges and universities. NASA Ames and NAFEO have established a partnership to explore bringing Historically Black Colleges and Universities (HBCUs) and Minority Serving Institutions (MSIs) faculty, researchers, and students to Silicon Valley as part of the planned NASA Research Park. A number of the Research Park agreements include emphasis on female and minority workforce development. Bringing minority students and faculty from minority-serving institutions around the nation through the NAFEO partnership will immediately expand the NASA Research Park into a national educational resource, while connecting those universities to Silicon Valley. "Developing a presence in Silicon Valley is important to our mission of 'Keeping the Doors of Opportunity Open,'" said NAFEO CEO and President Dr. Henry Ponder in a November 1, 2000 press release. "This research partnership with Ames at the NASA Research Park is an outstanding opportunity to bring faculty and students

from our 118 member minority institutions to where the action is for the New Economy," Ponder said.
- *Carnegie Mellon University*, Pittsburgh, Pennsylvania: Researchers from this university would complement some of the key technology strengths of Ames, especially information technology, high-reliability computing, and robotics. On December 11, 2000, Ames and Carnegie Mellon announced the formation of a new High Dependability Computing Consortium (HDCC), whose mission is to eliminate failures in computing systems critical to the welfare of society. Twelve information technology companies have agreed to work with Carnegie Mellon and NASA on the consortium and its agenda to promote and conduct research enabling the development of highly dependable, affordable software systems. The consortium's industry partners include Adobe Systems, Inc., Compaq Computer Corp., Hewlett-Packard Corp., IBM Corp., ILOG, Inc., Marimba, Inc., Microsoft Corp., Novell, Inc., SGI, Inc., Siebel Systems, Inc., Sybase, Inc., and Sun Microsystems, Inc. The High Dependability Computing Consortium represents the first concrete step in Carnegie Mellon's plan to develop a presence in Silicon Valley, which will include a branch campus for education and research programs. "Carnegie Mellon has a long history of building practical computing systems and is recognized for its expertise in software engineering," said Carnegie Mellon President Dr. Jared L. Cohon at the press conference announcing the HDCC. "We have an innovative faculty that excels in cross-disciplinary research. The university has played a lead role in forming this HDCC consortium, and along with the branch campus, we will showcase our research and educational offerings in Silicon Valley, the information technology capital of the world," Cohon said.
- *Lockheed Martin*: Ames has entered into a memorandum of understanding with Lockheed Martin's space operations company (based in Houston) to enable opportunities in new research. The first major collaborative project will be to construct a Laboratory for Advanced Space Research (tentative plans call for naming the laboratory after Carl Sagan and for establishing a Sagan library on the site). Lockheed Martin would be responsible for developing a critical portion of the infrastructure, upgrading older buildings, and building new office and laboratory space for sublease to tenants whose goals are consistent with NASA's mission. Plans are currently under development for construction to begin in summer 2001. Lockheed Martin already has extensive experience in collaborative ventures, including the lead role in the following consortia: Arctic Slope Regional Coalition, SERC, and the Martin Group.

Ames has discussed possible partnerships with other high-tech firms, including Oracle (information technology research), Raytheon (information technology, remote sensing, air traffic capacity, education), Sun Microsystems (information technology), AMD (nanotechnology), TRW (nanotechnology, IT, space sensors), Intel (information technology), and Computer Sciences Corporation (ATM technologies). All have expressed an interest in leasing space at Ames, in working with Ames researchers, and in developing either precompetitive or commercial technologies. When new infrastructure and rental space is available, Ames can begin establishing relationships with industry partners. Partners will participate in a "resident council" to address common issues for local resolution.

Educational Goals

One of NASA's explicit missions is education and outreach. Ames, because of its large holding of land, has the opportunity to create a unique community of researchers, students, and educators to support the dual objectives of research and education.

With this objective, one goal for Ames is to enhance the education of its own workforce, in multiple ways. One plan is to arrange joint appointments for staff members with university partners. Another is to form joint research teams on campus with researchers from other institutions. A third is to bring graduate students and postdocs with the latest training to the Ames campus.

The 350,000-square-foot Hangar One and surrounding areas at Ames will be converted into a world-class education facility called the California Air and Space Center. This center will be a 501(C)(3) corporation with six directors, including astronaut Sally Ride, director James Cameron, and appointees of the cities of Mountain View and Sunnyvale. The directors will establish a vision statement, create a business plan, and work with the support of NASA resources. In addition to interior space large enough to house numerous life-sized space vehicles, Hangar One will be home to the Teacher Institute.

Other educational projects include the nonprofit Computer History Museum Center, The Research Institute for Advanced Computation, and an expanded small-business incubator. Financial analysis has indicated that these educational projects will be able to operate on a self-funding basis.

A Regional Vision

Under the Space Act, NASA's use of federal land must be both appropriate to NASA's missions and responsive to the surrounding region. Ames, traditionally a fenced, stand-alone enclave, plans to open its gates to the community and enter more closely into the concerns of the surrounding region:

- *The environment*: Ames is adjacent to extensive wetlands, much of which is being used as commercial salt basins. The state plans to purchase these wetlands and return them to their original condition and Ames intends to

make its activities consistent with this conservation effort. In addition, it is working to complete the last link in a San Francisco-Oakland walking trail, The Bay Trail, across the northern edge of its property.
- *Transportation*: With the restoration of the wetlands is a plan to initiate new ferry service across the Bay to ease traffic congestion, with the terminal sited at Ames. In addition, Santa Clara County has extended light rail by building a terminal at Ames to facilitate travel through Santa Clara to South San Jose and to the Caltrans railroad station in Mountain View.
- *Regional development*: Ames has met with numerous potential partners and local neighbors (the Silicon Valley Manufacturing Group, the Bay Area Economic Forum, Joint Venture Silicon Valley, focus groups, local schools and universities) about the regional effects of the partnership plan. Responses from all groups have been positive.
- *Local governments*: The city councils of Mountain View, Sunnyvale, and other surrounding communities have discussed and endorsed the plan, and agreed to improve transportation and other vital infrastructure.
- *Access and housing*: Ames will by stages open its gates and remove security barriers from the research (western) portion of the complex. A new bridge may be built by Mountain View to provide access to undeveloped land that will form part of the research area. The center may be allowed use some of the 800 units of military housing to provide basic, inexpensive housing for visiting researchers and students.
- *Historic preservation*: A number of former military buildings, some of them located in a historic zone, are structurally sound but require upgrades to meet modern seismic and accessibility codes. The center plans to lease some of them to partners who will bring them up to code in lieu of rental fees. The first such tenant has entered negotiations. Other buildings may be removed and replaced with new buildings of architectural styles that are consistent with the historic district.

Proposed Schedule

A number of activities have already begun: completion of the infrastructure assessment, traffic studies, protection of historic resources, and initial massing studies and site plans. The center plans to invite potential partners to a planning session to align the objectives of Ames with those of its potential partners. Construction and gate changes may begin as early as the end of 2000.

In June 2000, NASA Ames announced its Environmental Impact Statement that includes a comprehensive plan for development of the entire 2,000-acre site. The EIS process under the National Environmental Policy Act (NEPA) will take up to 18 months. NASA Ames plans to reach a Record of Decision in late 2001 that will allow a build-out of an additional approximately 3 million square feet of space for the Research Park.

Summary

In summary, Mr. Berry highlighted these features of the Ames proposal:

- The proposal has the potential to establish a new world-class R&D campus that can benefit not only NASA but also the region and the nation; and
- Ames administrators believe that their plan will allow the agency to leverage NASA resources, the capacities of Ames Research Center, and the lands inherited by the center from the military for the benefit of NASA's mission.

The objectives of the strategic plan include improved scientific research and technology development; enhancement of NASA's education and outreach programs; productive partnerships with universities and private companies; and stronger workforces both at Ames and in the region.

PARTNERING WITH THE UNIVERSITY OF CALIFORNIA AT SANTA CRUZ

M.R.C. Greenwood
University of California at Santa Cruz

Dr. Greenwood's presentation explored why the University of California, and the Santa Cruz campus in particular, are interested in the Ames program. She emphasized two topics: research partnerships and the opportunity to "really deal with some of the issues around the digital divide"—that is, providing better technology education for the present and future workforces. Noting the shortage of human resources in engineering, she said that UC Santa Cruz is committed to a 50 percent increase in engineering graduates. The school has opened "the first 21st-century engineering school" and its plans for growth involve its partnership with the Ames program.

Some Strengths of UC

Dr. Greenwood said that the value of the University of California as a partner lay in its research strengths. UC is widely perceived as the number-one public research university in the world. The system already has 159 research grants with NASA, 34 of which are on the Santa Cruz campus. UC has a reputation for innovation and leadership, she said, and for making an "enormously positive" economic impact in California and the nation generally. Six of UC's nine campuses have already launched successful research parks, as well as partnerships with national laboratories and other organizations, such as Lockheed Martin. Of the $13.6 billion UC budget, over $2 billion is spent on research.

UC Santa Cruz, founded in 1965, is one of the younger, faster-growing campuses, said Dr. Greenwood. It plans to hire some 600 new faculty in the next decade, providing many opportunities for joint appointments and new programs. Its young engineering school has grown from several hundred students three years ago to over 900 at present, and already ranks number 15 nationally among public research universities.

Its major engineering research and teaching programs overlap closely with the subfields planned for expansion at Ames. In the partnership, it would emphasize the "mission triangle" fields of NASA (information technology, nanotechnology, biotechnology), as well as planetary sciences and astrobiology. It would also seek ways to address major issues of the "digital divide," and emphasize both K-12 and teacher education.

Outreach and Partnerships

The University of California at Santa Cruz has an extensive outreach to the Valley, through, for example, its Lifelong Learning program, which now enrolls some 52,000 people with activities designed to respond to existing companies and their workforce needs. The university also collaborates in outreach with the Foothill-De Anza Community College District and San Jose State system on high-tech educational initiatives.

The UC campuses have a good record of working with industries, especially in the field of biotechnology, which Dr. Greenwood described as "essentially a California industry." One of three biotech firms in the nation is within 35 miles of a UC campus, she said; one in four was started by UC scientists, including three of the world's largest. The industry now supports over 60,000 jobs in California.

The multi-campus structure of UC has led to considerable experience in partnerships. UC Santa Cruz, for example, serves as system-wide headquarters for the astronomy program, and has launched an S&T center in adaptive optics in partnership with four other University of California campuses, other universities, and 23 businesses. Santa Cruz would be an experienced anchor tenant for Ames, said Dr. Greenwood, and a portal to facilitate the identification of talent throughout the UC system. UCSC has also formed partnerships with other entities of business and government.

UC Santa Cruz has discussed a role as the lead university for the new Ames astrobiology laboratory, for which it would assemble teams in astrobiology, information technology, and nanotechnology. Strength in planetary sciences is already present, as is a major multi-campus research unit in geophysics and planetary physics. In addition, UC Santa Cruz and Ames would collaborate in the fields of remote sensing, data visualization, Mars missions, and space biology.

Strength in Education

A particular strength and interest of UC Santa Cruz is K-12 education. The University of California system, as a whole, has been given a "huge" role by the governor to improve teacher education and retention, and to hold algebra and science institutes and summer programs for teachers. UC Santa Cruz plans to combine aspects of these objectives with those of Ames' Teacher Institute for considerable impact in the region. The state legislature is motivated to address issues of the digital society, including early identification and inspiration of talented students. Although K-12 education is not a primary objective of the Space Act, said Dr. Greenwood, the educational partnership between Ames and UC Santa Cruz would be a good model for government agencies.

As lead university in the UC system for K-12 outreach, UC Santa Cruz has formed an educational collaborative with San Jose State University, the Foothills/De Anza colleges, local organizations, and UC resources. The model combines outreach, articulation, and programming, focusing on digital divide issues. For higher education, one expected outcome of the collaborative are joint doctorate degrees in science and mathematics education and in engineering with San Jose State and NASA.

The partnership between UC Santa Cruz and NASA began nearly two years ago, and has recently accelerated. The role of UC is to furnish planning money for the regional center, and perhaps for the educational collaborative as well. A UC academic task force will visit Ames in summer 2000, and siting/real estate teams are working with Lockheed Martin (with whom UC has long experience collaborating in DoE labs). The president of UC Santa Cruz has talked with leaders in Silicon Valley about partnerships, and a letter of intent with NASA is in progress.

In conclusion, Dr. Greenwood said that a strength of the partnership between NASA and UC Santa Cruz is their close alignment of objectives, in research, education, and outreach. The opportunities for joint appointments, joint projects, and workforce enhancements are extensions of ongoing efforts and goals for both institutions. For UC Santa Cruz, the practical objective is to provide Ames with a portal to the UC system; the larger goal is to create a nationally recognized model for collaboration between education, government, and industry.

THE ROLE OF LOCKHEED MARTIN

William Ballhaus
Lockheed Martin Corporation

Dr. Ballhaus, a former director of NASA Ames Research Center, praised the new Ames strategic plan, saying it could provide a "wellspring of innovation" that would benefit not only NASA and its partners but also the entire science and engineering enterprise. He said that Lockheed Martin (LM) has been collaborating on this strategic plan for some time and is eager to participate as a partner.

He summarized the mission of Lockheed Martin as "systems integration and technology," and said that this mission makes an excellent fit with the goals described by Drs. Venneri and McDonald. Lockheed Martin already has extensive working relationships with NASA in which it contributes strategic planning, program management, and laboratory support. The corporation employs a total of about 50,000 scientists and engineers and supports over 1,000 internal R&D projects and a variety of government contracts in aeronautics, space systems, information technology, electronics, technical services, and others.

In the 1990s, when Lockheed Martin took its present form, its leaders decided against maintaining a central research laboratory. They recognized that the scope of LM technology could not be covered by a single lab. Instead, the corporation reached out to form partnerships with a broad range of external sources, including Sandia National Laboratories (which LM operates), General Electric, Oak Ridge National Laboratory, and over 100 colleges and universities.

Partnerships and Clusters

Dr. Ballhaus described a recent meeting of the Council on Competitiveness which he attended in Washington. He said that the Council focused on three areas, all of which were vital elements of the Ames plan:

1) a technologically competent workforce;
2) enhanced government funding of basic research; and
3) regional alliances and incubators.

The meeting heard that, in the 1990s, businesses faced up to problems that impeded productivity and became more efficient. In 2000 and beyond, the next advances will be provided by innovation, to which regional clusters are critically important, including a supplier base, expert financiers, and supportive educational and governmental institutions.

Dr. Ballhaus suggested that such a cluster could take shape around Ames and the partnerships it forms. As one such partner, Lockheed Martin's goals would be to serve the customer's needs and to help reach the goals described by NASA. One reason that Lockheed Martin is an appropriate partner, he said, its is expertise in recruiting and managing human resources. LM hires 2,000 people every year (65 percent of them in electrical engineering and computer sciences). Another reason is that its focus on systems integration brings wide experience in collaboration.

He noted that the growing importance of collaborative research is illustrated by a recent study for the Air Force. The study recommended a science and technology strategy called GOCA—"government owned/collaborator assisted"—as a way to ensure a continuing stream of fresh ideas and talent.

He then cited four criteria he had used as director of Ames to assess the vitality of a research center: 1) the strength of its mission; 2) the quality of its workforce; 3) the uniqueness of its facilities; and 4) the strength of its interactions with academia, government, and industry. He said that Ames as a facility is strong in each area, and that the new strategic plan could establish a "new model for doing business with NASA."

Collaboration with Ames and Academic Partners

Within this model, one of LM's goals would be to establish and direct a Research Initiative Fund to support new research programs. Profits from such programs would be held in an escrow account, and the use of profits would be decided jointly by NASA, LM, and UC Santa Cruz. Potential uses would include innovative research programs, academic fellowships and joint educational projects, K-12 initiatives, and various vehicles to further the mission of NASA. Lockheed Martin would also be responsible for developing the new air and space museum and the research and educational environment that supports Ames' mission.

Specific areas of research collaboration with Ames would include astrobiology, information technology (LM is a leader in government and commercial IT system design and implementation), nanotechnology (taking advantage of leading-edge research at Sandia National Laboratory, managed by LM), life and microgravity sciences, and aeronautical and space technology. To strengthen this collaboration, LM would also join with Ames in workforce enhancement through joint appointments and internships, access to graduate students, postdocs, and future employees, and on-site continuing education.

LM would also collaborate on several fronts with Ames' major academic partners, UC Santa Cruz and Carnegie Mellon University. First, university professors would play an educational role in the laboratories where LM researchers work. In addition, LM and its academic partners would collaborate on both scientific and educational programs, including joint R&D projects, advanced education for LM employees and other Center tenants, shared leadership in the Research Initiative Fund, and technology transfer.

Experience with Partnerships

Lockheed Martin has already experienced aspects of the Ames plan in its role of manager of Sandia National Laboratories, a Department of Energy facility in New Mexico. With its management contract comes a commitment to investing in the local community. An important aspect of this investment is the Technology Ventures Corp., a 501(c)(3) corporation founded in 1993 to facilitate commercialization of new technology, commercialize technology developed at the lab, and create new businesses and jobs in New Mexico. Some early results include the following:

- strategic alliances with 25 partners;
- establishment of a 200-acre science and technology park;
- 7 tenant companies employing more than 225 people;
- investments in client companies totaling $166 million;
- 36 business formations, 27 business expansions, 16 licenses, 7 CRADAs; and
- more than 2,000 jobs created.

Similarly, a 250-acre multipurpose industrial office park has been created at Oak Ridge. Park management is committed to invest 10 percent of profits from park activities in the local Oak Ridge community. This is done through educational institutions, entrepreneurial ventures, United Way, and other means. The Oak Ridge initiative:

- supports a model technology transfer program, which makes available the latest equipment and staff expertise. This has resulted in more than 250 licenses of Oak Ridge technologies, hundreds of CRADAs, a small business incubator, and an investment company to supply seed capital to technologies developed by DoE. It also
- supports Technology 2020, a public-private partnership that leverages IT resources to create an entrepreneurial environment, develop a high-speed information infrastructure, and establish a pipeline of qualified IT professionals; and
- contributes millions of dollars to education, corporate matching gifts, and education scholarships.

Dr. Ballhaus closed his talk by congratulating the Ames leaders on their plan, and citing the value of linkages and collaborative research to Lockheed Martin. "We have common goals with NASA and the other partners," he said. "A shared R&D center will benefit us all. We are very much focused on linkages into the wellsprings of intellectual activity and new technology in areas that affect our mission."

THE ROLE OF CARNEGIE MELLON

Duane Adams and James Morris
Carnegie Mellon University

Moderator Patrick Windham introduced Dr. Adams as a former director of DARPA who has had considerable experience with government-industry partnerships. Dr. Adams began by praising the plan for a research park at Ames, citing the advantages of collaborative programs over programs done in separate organizations. He suggested that the Ames design include explicit strategies to induce interaction, including opportunities for communal dining, on-site lodging, and round-the-clock access to research facilities. "It should be a place where people from the research community want to come, for employment or visits or sabbaticals, to take advantage of the environment."

A Commitment to Collaboration

Dr. Adams said that Carnegie Mellon University (CMU), of Pittsburgh, Pennsylvania, was committed to working as a partner with NASA, with other universities, and with corporate partners in forming and operating a "world-class, shared-use research park." The focus of CMU's activities will be research, especially long-term, fundamental investigations in the areas of robotics, software engineering, and other aspects of information technology. He affirmed that education is most effective in the context of research, and for this reason he advocated the participation of CMU students, as well as faculty (typically during sabbatical leave from campus), as part of the local research community. He cautioned that structuring the space and buildings in ways that truly facilitate collaboration will take "a really collaborative effort among all major partners."

CMU has managed a program in robotics for 20 years, initiated with flexible startup grants from Westinghouse. The sophistication of this Federally Funded Research and Development Center (FFRDC) has increased over its lifespan. An initial focus on manufacture, inspection of circuit boards, welding, and forging has evolved to include robot fields, health care robots, and exploration robots. NASA has funded the National Robotic Engineering Consortium, along with about 20 commercial partners, including Caterpillar, New Holland, and Ford, with the goal of transferring technology developed at CMU, NASA, and elsewhere. The robotics institute includes an educational program with Ph.D. and master's degrees.

The software engineering program began 15 years ago in response to defense needs. About a dozen years ago the program set up a computer emergency response team for DARPA to counter an internet threat. Today the program has matured into a "911" system for cyberattacks. It, too, is an educational program, offering a master's in software engineering.

Possible Lessons for Ames

Through such collaborations, CMU has learned valuable lessons that may be useful at Ames:

- Institutional support from the top is essential. At CMU, a new program is

usually initiated by a principal investigator with a proposal, but its success depends on the active support of institutional leadership, "all the way up the entire chain," from the deans and provost to the president and board of trustees.
- In both robotics and software engineering, flexibility and continuous innovation are essential to keep programs vigorous. "The projects you start out with are not necessarily the ones you end up with after a few years."
- Industry relationships are essential to productive collaborations, but they must be nurtured patiently over a long period. "It is important to deal not only with researchers, but right up to the CEO and other key leaders."
- Success will depend on the Center's ability to attract the very best researchers. This ability can be enhanced by permitting and encouraging companies to co-locate on the site and by providing incubator services for new and small businesses.
- Even within individual institutions, collaborations among departments can spur innovation. Mutual interests between the computer sciences and business studies programs at CMU have resulted in a master's program in e-commerce.
- Thanks to modern electronic communication, it will be possible to perform tightly-coupled research programs in separate locations (Ames and CMU campus).

As a partner at Ames, CMU would have the following objectives: to perform collaborative research with NASA, other universities, and Silicon Valley companies; to participate in selected educational programs to benefit NASA and private partners; and to assist in the commercialization of technology by various means, including hosting and incubation services.

"At Ames, we need to share ideas and really work together," Dr. Adams concluded. "We have to attract enough researchers and provide incentives for them to do more than just research. They must have opportunities to commercialize their work and participate in ventures. These are important incentives."

Dr. Morris continued the discussion of CMU's role by suggesting that the time is right to launch a major, long-term effort in "dependable computing." The time will soon come, he said, when the "euphoric era of computing will end" and users will no longer tolerate bugs and failures. He said that society's attitude toward computing, which began with an atmosphere of almost euphoric acceptance, will rapidly evolve into the same demand for reliability experienced with the automobile several decades ago.

Dr. Morris said that most computer systems today are assembled from commercial, off-the-shelf components that may not always work. To make them dependable, methods and theories of systems integration are needed that have not yet been developed. Systems must also be protected against attack from malicious forces.

Embedded systems are already essential to the operation of devices as mundane as water heaters and telephones, and they will become more so. They are not traditionally considered critical systems, but the day will soon come when consumers demand that embedded devices must be fail-safe, along with the goods deployed in international markets.

In particular, Dr. Morris suggested that the multi-decade, "dependable computing" program be coordinated by a panel called the High Dependability Computing Consortium. This consortium would seek to provide a sound theoretical, scientific, and technological basis for construction of safe, secure computing systems. More specific goals would be to protect the public, protect the consumer, preserve competition in the computer industry, and promote national security. It would seek to develop high-assurance computing elements for every sector, including transportation, medical systems, consumer products, and national security.

Members of this consortium, who have conducted preliminary discussions, would be based at Ames and likely would be composed of the following participants: 1) universities, including CMU, University of California, MIT, University of Washington; 2) other government agencies in addition to NASA; and 3) private firms, including Sun Microsystems, Adobe, Compaq, Microsoft, Marimba, ILOG, SGI, Siebel, Novell, IBM, and Hewlett-Packard.

Current Needs in Robotics and Computing

Dr. Morris outlined a list of additional activities that would be appropriate for partnerships at Ames:

- *Software engineering*: Software engineering—the partner of hardware engineering and systems engineering—must be developed in both its research and educational aspects. Not enough is known about using formal methods to check systems. Formal methods have worked well to design hardware and chip systems, partly because they are smaller systems with tighter specifications, but they have not been pushed into the software area.
- *Software clusters*: In addition to software engineering research, major areas of software education are needed, as distinct from computer science education. Many software engineers are people who were engineers and then learned either software or computer science. True software engineering curricula are needed. These might resemble medical curricula, where clinical doctors understand both the underlying biomedicine and also enjoy practicing health care. Software educators would truly understand how software systems are built and enjoy teaching. People in Silicon Valley who have built the systems (including some of the 2,500 CMU alumni working there) could teach software engineering in a completely new way.

- *Human-computer systems*: A crucial area of research is human-computer "systems." Some 80 percent of airline accidents are attributed to pilot error. Similarly, computer system failures are in reality often programmer or user errors. On a psychological level, it is essential to better understand humans as fallible "parts" of computer systems. The new Human Computer Interaction Institute at CMU is devoted to this subject.
- *Basic computer science*: A great deal of basic computer science remains to be done. Dr. Morris gave the example of a colleague who recently invented a concept called proof-carrying code that allows one to test a piece of code from anywhere in the world in the assurance that it will not disrupt a computer system.

Robotics for Planetary Exploration and Life-Seeking Missions

Dr. Adams returned to explain that the development of more reliable and autonomous computing systems would directly promote the development of exploration robots, which must operate in the most extreme and remote physical settings, both on Earth and on other planets. Immediate goals of "space robotics" would include planetary exploration and life-seeking experiments in extreme environments.

Life-seeking experiments depend on the kind of collaborative research envisioned by the Ames strategic plan. In this case, roboticists and biologists would collaborate to develop life-seeking detection instrumentation; non-intrusive, minimal-contamination robotic search techniques; and autonomous means to discover and classify chemical and life forms.

Likewise, planetary exploration is a collaborative venture requiring diverse expertise. One of the challenges of planetary exploration is to seek out and use terrestrial venues that resemble the environments found on other worlds. For example, Haughton Crater on the dry plains of Antarctica provides an analog for the Martian permafrost environment; polar volcanoes and fumaroles provide an analog for Martian volcanoes; Lake Vostok provides an analog for the polar sub-ice environment; and the Chernobyl reactor provides an analog for a high-radiation environment. Researchers at Ames can monitor the use of systems here on Earth before sending them on expensive missions in space.

Planetary global explorers would be expected to have decades of operational life, thousands of kilometers of range, and sufficient autonomy to require minimal human intervention. They would be asked to perform comprehensive regional scientific surveys in the most extreme environment. An example that has already been studied in some detail by researchers at CMU and NASA is the Victoria robot system for surveying the surface of the Moon. Victoria would circumnavigate the lunar south pole, where "she" could receive solar energy at all times by traveling in a sun-synchronous route in a direction opposite the Moon's rotation. The route would be designed so that the robot would complete one polar cycle each month, surveying the geological environment and relaying data continuously to Earth. More generally, the Victoria project would provide a focus and stimulus for research on robust mechanical, computational, and electrical components, and in autonomous navigation and fault detection equipment.

Dr. Adams concluded with a few examples of the capabilities of some of CMU's exploration robots:

- A robot named Nomad has been launched in the desert of Chile and guided on a 200-km journey by remote control from Pittsburgh. Young students came to the science center and learned that they could control a robot in another continent, in a barren environment that might resemble that of another planet. The same robot found five meteorites autonomously in Antarctica.
- A robot called Dante—a "legged walker"—descended into the volcano of Mt. Spur in Alaska.

Various commercial vehicles fitted with sensors and control systems have harvested hay, mined ore underground, and hauled crushed ore from a mine.

DISCUSSANTS

Robert Wilson
University of Texas at Austin

Potentially Pathbreaking

Dr. Wilson began by saying "I find this a truly extraordinary proposal" which "seems to have the potential for a path breaking arrangement." He noted in particular the scope of the proposal, the range of resources available, and the stature of the institutions involved. He added that he did not know of any other government agency with such an extensive collaboration that involved so many functions, including basic research, technology transfer, land and property management, housing development, environmental protection, public information, and education.

But Points to Consider

Turning to his role as discussant, Dr. Wilson recognized that the Ames' proposal to broaden its range of activities and to mobilize nongovernmental resources for its principal mission represents a response to Congressional action and contains an element of "reinventing government." Wilson identified several potential challenges for implementation of the plan. The first he described as "some disconnection" between the principal missions of Ames, which are technology- and research-oriented, and the broader objectives of the new strategic plan. The plan reflects "more than a subtle shift in mission," he said. The long list of partnerships entailed in the plan would require "an enormous amount of institutional energy," because no partnership is easy—even if the partner is familiar and shares common goals.

He also raised the difficulty of defining and measuring success for these new objectives. For example, "Ames as landlord" will be concerned with recovering costs through rental income. On the other hand, "Ames as educator" will face the challenges of designing training programs and evaluating their impact. The education of students is a powerful means of transferring technology, but evaluating the effectiveness of this transfer is itself challenging. He said that the program is ambitious and broad, and many of its elements will have to be evaluated individually as well as in terms of their contribution to the core missions of Ames.

Given the importance of the expanded Ames role in land management in one of the country's hottest real estate markets, the difficulties of regional governance systems will likely emerge. Ames has incorporated important local governmental jurisdictions in the development of the plan, but the presence of Ames, as a federal agency, in local planning and land management will create a need for creative, intergovernmental, and interorganizational systems of governance. Just as the Ames activities will affect adjacent communities, adjacent communities will also have an impact on the project.

Finally, Dr. Wilson suggested that one of the reasons Ames has formulated this plan is "to remain competitive." He pointed out that many universities and research centers are focusing on these new technologies. Given the multiple objectives of the plan, how will Ames know if this endeavor has been successful?

Edward Penhoet
University of California at Berkeley and Chiron Corporation

Dr. Penhoet offered two suggestions about the Ames plan for partnerships. First, it is important to make sure that collaboration means more than having a collection of new buildings on the site. He suggested that teamwork is powerfully stimulated when partners share major pieces of equipment, such as the advanced light source at Lawrence Livermore that is used by people from Berkeley and other institutions.

Second, he suggested taking the bold step of housing people from three or four institutions in the same building—to further encourage and even force interaction among researchers.

QUESTIONS & COMMENTS

Mr. Berry responded that a portion of Ames funding will be committed to outfitting the new space science laboratory for collaborative research. Ames has revised its planning several times in an effort to make the site attractive to people from different disciplines.

In response to Dr. Wilson, Mr. Berry remarked that the only organizations with which Ames competes are the other NASA centers. "Competition is one element of how we get authority for programs, and we are very much looking toward the long-term competitiveness of Ames within the NASA system," he said. "At the end of the day, my vision is, what have we accomplished in the way of research initiatives; have we really made the breakthroughs and accomplishments?

"We want a successful research park, joint degree programs, and new students going into the market," he continued. "And we want it to be economically viable. But the real objective is to enable a new future for NASA. If new things flow from this activity that allow NASA to do breakthrough missions 10 or 20 years from now, that is our metric for success."

Dr. McDonald added, "Our metric of success is how much we have enabled NASA to achieve the very lofty goals the administrator and the Congress and the President have set. Also, the educational activity with our various partners will have a vital synergism, and we will achieve important mentoring of the next generation of NASA employees and technologists. We have to compete with industry for the best and the brightest, so we have a profound interest in the educational process."

Dr. Wessner raised the question of financing, asking whether there would be pressure to ask for additional federal funding to support the intense level of activity. Mr. Berry replied that Ames plans to use appropriated funds to facilitate creation of the Research Park, but that operations and programs would be supported by leasing income and revenues from partnerships. He added that income generation would be possible primarily because of the value of Ames' location to prospective partners.

Dr. Greenwood said that UC Santa Cruz had developed a successful financing model in its outreach partnerships in Silicon Valley. Given the student presence, the university has the capacity to leverage state funds by indirect cost recovery, along with some fund-raising, to provide classroom and other facilities. She suggested that this model could be extended to Ames.

As a final comment, Burton McMurtry expressed his concern that the scarcity of affordable housing in Silicon Valley might threaten the project's viability by undermining the ability to bring in new employees necessary for the undertaking.

Dr. Morris was asked about a plan at Carnegie Mellon to create a four-year Ph.D. program that would be divided evenly between time on campus and time in an industrial lab. He answered that despite many reasons in favor of trying this new paradigm, the traditions of the campus have restrained execution of the plan. Ames would afford a fresh opportunity to develop the idea.

Dr. Penhoet added that Ames would provide an excellent environment for other new learning paradigms, such as breaking the exclusively "sequestered," on-campus setting in favor of internships and intensive off-campus learning experiences.

Dr. Wilson asked for more detail about how the research fund mentioned by Dr. Ballhaus would be funded. Dr. Ballhaus replied that the initiative would be funded from the profits of partnerships, revenues from leases, and other rental fees. The fund would then be administered jointly by the partners.

Mr. Berry added that NASA plans to update its research priorities annually, provide these priorities to its partners, and expect its partners to reinvest in research in ways consistent with those priorities.

Panel IV:
SBIR Initiatives and Mission Objectives

Moderator:
Burton McMurtry
Technology Venture Investors

IN-Q-TEL: A "NONPROFIT VENTURE CAPITAL FUND"

Gilman G. Louie

In-Q-Tel

In-Q-Tel, which Mr. Louie described as a "nonprofit venture capital fund," was established by the Central Intelligence Agency (CIA) to promote the private development of new technologies that might meet basic information needs of the Agency. He offered a brief description of his company as one model that might be of interest to NASA in its own efforts to stimulate technology development.

The CIA was established in 1947 when President Truman realized that the government probably possessed enough information to have predicted the attack on Pearl Harbor on December 7, 1941—but that the information was too dispersed to allow proper interpretation. The CIA was charged with centralizing and summarizing such vital information. Today, however, even the government's primary information agency is overwhelmed by difficulties in centralizing, mining, and summarizing the vast amounts of data gathered by satellites, computers, and other modern devices.

Capturing Fast-Moving Technologies

In-Q-Tel[3] was set up on a dual premise: 1) Development of today's information technologies is driven largely by fast-moving young "dot-com" companies;

---

[3] The name is derived from <u>I</u>nformation, <u>Tel</u>ecommunications, and <u>Q</u>: a character in the Ian Fleming series of James Bond spy novels and films.

and 2) in order to tap into those technologies, the CIA needs to establish communication links with those new firms. The old language of RFPs and contracts, said Mr. Louie, while appropriate for many kinds of acquisitions, is less useful for obtaining useful new technologies because of the speed with which they advance. By the time the CIA uses traditional means to identify a need, process it, acquire it, and adopt it, six months to two or more years may have passed, rendering the technology obsolete.

The goal of In-Q-Tel as a nonprofit venture capital fund is to roll any profit back into R&D. This approach is not appropriate for classified projects, but for unclassified challenges the goal of the fund is to find new technologies in the marketplace and move them into government more efficiently and quickly.

Mr. Louie, who heads In-Q-Tel, began his own career by founding a small company in Silicon Valley. He counts his experience and contacts there as essential in finding new technologies. In practice, In-Q-Tel monitors technological developments at universities, FFRDCs, private firms, and venture funds. The last are especially important because they provide advance indicators of where the marketplace is going. When In-Q-Tel finds a company doing work that is relevant to the CIA's needs, it can offer the company "real-world" feedback. Half the staff works in Silicon Valley (or other high-tech regions) where they search for promising technologies that might fit the Agency's needs. When they find one, they invest in it, help it along the path toward commercialization, and integrate it into the Agency.

Part Incubator, Part Venture Capitalist

Part of In-Q-Tel's work is to act as an incubator. Technical members of the staff help new companies through technical hurdles, and business people help companies create business plans and raise money. In-Q-Tel also functions as a venture capitalist: last year the company allocated over $10 million of a $34 million budget to venture capital for direct equity-style investments; much of the rest of the money is spent on "hybrids," agreements with elements of both contracting and equity conversion. Sometimes In-Q-Tel contracts with companies with the stipulation that the invested money be converted to equity if In-Q-Tel shows the innovators how to take their new technology not only into the CIA but also to the commercial market.

The decision of the CIA to work with In-Q-Tel sparked a debate about open standards. On one side was the desire to keep new technologies closed and secure, where no one could test or tamper with them. The winning side, however, argued that a truly durable technology is one that can live on the internet, where any weakness can be exposed early by hackers and then remedied before it is released. Open standards also help stimulate competition in the marketplace, bringing improved technology.

Motivation Through Equity

Mr. Louie explained why a contract is less effective for developing new technologies, citing two reasons. With a contract, he said, goals and metrics must be explained at the outset of an agreement. With new, fast-moving technologies, however, it is seldom possible to know the outcome of a project in advance. Second, part of a contractor's motive to succeed is to avoid the penalties built into the contract. With an investment or partnership, the contractor is motivated not only to succeed but to do its very best in the hopes of finding, patenting, and profiting from a new technology.

He offered an example of the effectiveness of this process. A company was developing a promising internet security technology supported by In-Q-Tel. The company called to ask if they could speed up the timetable by 50 percent—and also invest $10 million of their own money. The company's strategic interest was being served by getting to market quickly, and In-Q-Tel's interest was being served because they wanted the technology tested in public before it was moved to the CIA.

Mr. Louie concluded by emphasizing that finding new technologies was only part of the job. The other part is to constantly survey and review Agency requirements that might be addressed by those technologies. As prototypes are developed, the technology has to be converted into a form in which analysts and other Agency technicians can use it to their advantage.

QUESTIONS & COMMENTS

Dr. Wessner asked about the reactions of Congressional committees to In-Q-Tel's role. Mr. Louie replied that he had testified in the House before both the Appropriations and Intelligence committees. Both, he said, were concerned that the program did not become corporate welfare, and that it maintained a fair and level playing field for all potential partners. He said that the company is "living under a lot of scrutiny."

Kathy Behrens asked how the success of In-Q-Tel was being measured, and the time frame for success. Mr. Louie answered that the metrics for success were simple: are we able to accomplish the mission of finding valuable new technologies for the CIA that people are actually going to use? He said that they had 18-24 months to demonstrate success.

AN "ENTERPRISE FUND" FOR NASA

Robert L. Norwood
NASA

Dr. Norwood described the concept of a technology investment fund, or what might be called an "enterprise fund"[4] for NASA. The fund would be a new business organization, independent of government, with links both to NASA's technological resources and to the private investment community.

The goals of the enterprise fund would be 1) to identify NASA technologies (at Ames or other centers) with strong commercial potential, 2) to find business or corporate partners capable of developing those technologies and applying them to new products and services, and 3) to involve market savvy investors in the venture to carry them into the marketplace. Through the fund, NASA could be included in a development venture as a limited investment partner.

The enterprise fund would operate on return-on-investment principles, as set out by a professional management team and an industry-style board of directors. NASA would provide technical collaboration and management as needed.

Dr. Norwood said that the planning for this new fund has been driven by recent changes in the technology business environment to one that is "quick-reacting, high-tech, and market-based." He noted the explosive growth in venture funding, from about $6 billion in 1995 to about $45 billion in 1999, plus an additional $30 billion in funding provided by angel investors. Today's market, he said, is characterized by new entrepreneurial spirit and creativity, greater risk-taking and larger rewards. New markets and business/industry structures are continuously generated.

The new model is designed to take advantage of this new environment by expanding the opportunity to leverage and create technology-based ventures where NASA and business technology needs intersect. The fund would generate benefits in both directions. It would benefit NASA when efficiencies of the marketplace make new technologies (no matter where they are generated) available for purchase by NASA quickly and cheaply. The fund would benefit the investment community by providing readier access to useful and potentially profitable technologies developed by NASA.

Moving Technology into the Marketplace

Dr. Norwood said that NASA already creates several hundred partnerships each year with industry. What is new about the enterprise fund is that it would

---

[4] The term *enterprise fund* combines the notion of private enterprise with NASA's custom of describing its mission activities under four distinct "enterprises."

involve the investment community and leverage that community's "dynamic structure and agile operations."

"What we are trying to do," he said, "is use the best practices of identifying the market and leveraging the power of the marketplace to help NASA commercialize its technology."

The enterprise fund would take advantage of two existing resources. The first is the research base created by NASA's $100 million SBIR program authorized by Congress. This research spans 18 major technological areas. Every year about 70 companies graduate from phase 2 of the program, when their technology has reached the prototype stage and a business plan has been submitted. The new fund would essentially add a "phase 3," in which the technology would be developed further by a partnership. The second resource is NASA's in-house R&D program, which generates some $800 million worth of mission-relevant technology each year. This program includes five strategic areas, and 40 near-term development areas, all of which have a mandate to seek commercial partnerships. Major markets include rapid design tools, telecommunications, smart sensors, data mining, medical and environmental instrumentation, and information technology.

Dr. Norwood said that the entrepreneurial community has encountered difficulties in using the standard SBIR model to convert technology into commercial products. An enterprise fund would attempt to structure venture partnerships based on the NASA technologies in a new way. Rather than using traditional contracts for Phase III activities, the fund would create an investment agreement between NASA, a contractor, and the investment partners.

He said that there is risk in starting a company from an enterprise fund, as there is for any new venture, including NASA's own risk in providing the initial funding from existing sources. In addition, directors of the enterprise fund would have to contend with traditional private-sector suspicion of government involvement and the fear of controls that might delay or encumber innovation. But he suggested that the enterprise concept would soon allay any private-sector suspicions by demonstrating that it is "a new business organization independent of government controls and aligned with industry practices."

Dr. Norwood said that the authority to undertake such partnerships is already described in the Space Act under an "other transaction authority" that allows relationships with private firms. He added that the enterprise fund concept is not conceptually different from current licensing agreements to transfer knowledge to companies.

In conclusion, Dr. Norwood summarized the arguments in favor of the enterprise fund. For NASA, it would provide

- a profit center for innovations derived from NASA mission technologies and applied to commercial markets;
- opportunities to accelerate the development of mission-relevant technolo-

gies in the commercial marketplace and to acquire these technologies for use by NASA at lower cost; and
- a reduction in the business risks of technology transfer by involving the investment community and benefiting from its market expertise.

For the business and investment communities, the fund would provide

- opportunities to gain substantial returns on investment by leveraging NASA technologies;
- a capable partner in the acceleration of market-relevant technology; and
- direct linkage to the research strengths of NASA, which reduces technological risk for companies.

QUESTIONS & COMMENTS

Mr. Windham asked about the connection between the enterprise fund and the research park at Ames. Dr. Norwood replied that they would complement one another, and that the entrepreneurial activities of the fund would help move NASA technologies from all of its research centers into the marketplace.

Mr. Windham also asked whether the enterprise fund had advantages above the transaction authority that already exists and the Space Act authority that is similar to CRADAs. Dr. Norwood answered that the main purpose in designing the enterprise fund is to add the agility and drive of venture capitalists and other elements of the investment community.

A VENTURE CAPITAL PERSPECTIVE ON RESEARCH PARKS

Kathy Behrens
Robertson Stephens Investment Management

Dr. Behrens, who invests venture capital in high-tech firms for RS Investment Management, emphasized the "vast differences" between the world of scientific and engineering research and the world of venture capital. She questioned whether the two worlds could find sufficient overlap and alignment to form successful partnerships.

One difference concerns basic goals. Unlike most researchers in science and engineering, whose goal is to answer interesting questions and discover new products and processes, for traditional venture capitalists the primary goal of their work is financial return. Closely related to return is monetary compensation, which is based on the profitability of a company in which the venture capitalist has invested.

A second difference is that the main service of venture capital is not a product or a technology, but an application with concrete monetary value to the market.

A third difference is in the time scale of activities. Venture capitalists are compensated partly for their ability to "get there first"—to predict where the market is going and to be there when it arrives. When they sense a change in the market, they quickly—even instantaneously—make changes in their investments. Researchers, by contrast, must plan and develop support for their work gradually; government organizations traditionally make changes at a deliberate pace.

A variation on traditional venture capital—corporate venture capital—has far more in common with what Ames is trying to do. Goals are more strategic in nature, and time frames are longer. Certain features of corporate venture capital could provide a useful model for Ames.

Dr. Behrens suggested that Ames should not get into the "let's find a home for our technology" business. Venture capitalists prefer to finance people rather than specific, already-developed technology. She underscored the difficulty of taking a new technology—even a good one—and finding a market for it.

She said that Ames has a great geographical advantage in its Silicon Valley location, where the local economy is uniquely vibrant. At the same time, venture capital is "the most competitive business in the world today." Three years ago, she said, venture capitalists put $6 billion to work per year in new companies; by last year the figure had passed $50 billion in a single year. Ames would face a difficult challenge in learning the business, building up a network of contacts, communicating its story to the Valley, and learning to promote their technology. Hiring people who can "work in the Valley" is critical, she said, and those people must be listened to at Ames and have ample time to develop an Ames network.

QUESTIONS & COMMENTS

Dr. Luger questioned the focus of discussions on Silicon Valley, and asked whether other high-tech centers were not important as well. A number of participants agreed that technological development is now global, and that venture capitalists, like researchers, seek out partnerships all over the world. Dr. Behrens agreed, but emphasized that individual specialists in venture capital need to be physically close to their partners to operate effectively.

Panel V:
Ames as an Entrepreneurial Center: Opportunities and Challenges

Moderator:
Mark Myers
Xerox Corporation

COMMERCIALIZING TECHNOLOGY
Carolina Blake
Ames Research Center

Ms. Blake, chief of the Commercial Technology Office (CTO) at Ames, recalled the challenge issued by NASA Administrator Dan Goldin several years ago: "If Ames could find corporate partners willing to work on technologies that were both critical for NASA's mission and profitable for the companies, NASA would provide space for them to work at Ames." The response at Ames, she said, was to leverage what Ames already does by adding something new: an entrepreneurial center to expand the pool of technological resources through focused partnerships.

CTO Roles

The existing Commercial Technology Office at Ames has several roles: technology assessment (finding the right time to take a technology to market); marketing (throughout the U.S. and abroad); working with NASA's patent counsel to license intellectual property; and bringing companies into partnerships when the technology needs further development for patenting. The office is the focal point for business incubation at Ames and will be for the research park. Its mission, in the words of NASA headquarters, is "leveraging opportunities and partnerships with organizations outside of NASA in areas of emerging technologies."

New Procedures for New Firms

In response to this mandate, the current office will expand into a new Entrepreneurial Center. Among its goals are to devise ways of resolving common technological problems in ways that accelerate the spin-off of NASA technology and expand opportunities for NASA incubators, both at Ames and throughout the country. The office is trying to expand its resource pool to make this happen. It plans to focus on individual partnerships so that each has its own approval process, its own line of communication between NASA researchers and commercial partners, agreed beginning and end points, and milestones. The office will employ mini-CRADAs when possible, because of their flexibility, and manage the project once approved.

New Agreements

On March 13, 2000, the office signed an MOU with an internet consortium of companies. It is also working on a Space Act Agreement, a land lease agreement, and a programmatic agreement, which will call for each partner to put in two dollars for every dollar NASA invests.

Focus Areas

Initial areas of collaboration will probably include nanotechnology, biotechnology, and internet security.

Intellectual Property Issues

The office is working with NASA's legal counsel to resolve a number of problems regarding intellectual property. They will use existing authorities from the Space Act and the Stevenson-Wydler Act, and new models are being developed to allow some revenues to be invested back into new partnerships. The office is aware of the need to make government rules more flexible in working with industry.[5]

"As a technology transfer office," she said in conclusion, "we must protect the public investment, and our leadership. And so far, everything we plan makes use of authorities NASA already has. But we are also aware of the rigidities of government rules, and we have to examine these in light of the new realities of globalization."

---

[5] The STEP Board has launched a major review of U.S. intellectual property policy. See The National Research Council, *Intellectual Property Rights: How Far Should They be Extended: Report of a Workshop.* Washington, D.C.: National Academy Press, forthcoming. See also www.nationalacademies.org/ipr.

THE EXPERIENCE OF ONE START-UP COMPANY

Elizabeth Downing
3D Technology Laboratories

Dr. Downing is a founder of 3D Technology Laboratories, which is now four years old. She described an important distinction among start-up companies. Some have low technological risk and are able to attract angel or venture capital funding early on. 3D Technology Laboratories, however, is developing a technology with substantial risk and therefore requires a different funding path.

Developing a Technology

This technology, called cross-beam volumetric display, is a means of providing realistic and safe three-dimensional display. It employs a gated, two-frequency up-conversion and requires two elements to function: an active ion and a host medium in which the active ion can be doped and dispersed fairly uniformly. The ion itself has a number of energy levels, including infrared wavelengths and a different, excited-state wavelength. The result of the gated photonic excitation (as opposed to electronic excitation) is the emission of visible light. This allows the interception of two infrared laser beams. At the point of intersection, visible light is emitted, and scanning this light around the inside of an image chamber produces three-dimensional images.

The technology has been demonstrated and the company is now increasing its scale. It has a number of features. By addressing information in a volume, rather than a flat plane, it provides real stereo depth perception. There is no conflict between accommodation (the focusing action of the eye) and convergences (the angle between the eyes as they focus). This conflict is what causes headaches and nausea in stereo and shutter-glasses displays. Nor are glasses or headgear needed. It offers 360-degree, walk-around viewing of the data that's being displayed inside the image chamber; multiple viewers can see it simultaneously and interact with it. The images can be dynamic (refreshed at 30 hertz) and have the potential for multiple colors and opacity.

Another attribute of this technology is that the image chamber is a nonpixillated homogenous volume of material, which confers a manufacturing advantage. Traditional CRTs are pixilated, and liquid crystal displays are pixilated with wires and electrodes, so that conversions to 3-D would be complex. With the cross-beam system, information is addressed remotely: lasers are scanned remotely and modulated remotely. Then the material "does all the work." Unlike a laser, it poses no "eye fry" hazard because the radiation is incoherent.

The value of this technology has been recognized for decades, but potential developers—and venture financiers—were daunted by the issue of scaling. It was originally demonstrated on a very small sample, then scaled up to the order of a sugar cube, then to a cubic inch. 3D Technology Laboratories is currently attempting a chamber about 7 inches on a side, but the materials are difficult to make. Another issue is the size of the addressable data set, which requires rapid scanning; if scanning is too rapid, brightness is poor. The solution is to develop higher efficiency materials, which requires time and work. As Dr. Downing points out, however, the CRT was first presented in the 1920s as a dim, monochrome experience; 80 years later it is bright, colorful, and omnipresent.

A third area that requires more work is software. It is not a high-risk area, but little of it exists because there is no market yet.

Finding Support for Development

In her pursuit of funding to develop this technology, Dr. Downing has received little support from private firms or universities. Private firms saw that the technology had a long lead time and wanted rights to her intellectual property, her only real asset, before investing. Several partnerships with universities brought more difficulties than value to the company. In particular, she found the technology licensing offices to be demanding, even unreasonable, and one professor who was contracted to write software for the company attempted to profit from the software on his own. 3D now has a policy of avoiding partnerships with universities.

Instead, Dr. Downing has found support through a series of government grants, beginning with an SBIR contract when she was in graduate school. Soon after that a Phase 2 grant from NSF provided an essential foundation for development, and was followed by grants from DARPA, NIH, the Air Force, and more recently, the ATP.[6] In essence, "The company tries to mitigate the technology risk and make itself more appealing to outside investors by trying to solve some of the scanning and system architecture problems with government funding."

She added that venture capital firms were not interested in her technology as long as it required further development. At present, the company has a "fairly substantial" materials R&D program under way to improve efficiency and brightness and to improve the image chambers—as Dr. Downing says, to make them "bigger-brighter-cheaper-lighter." She feels that her work is important, even though it has taken 12 years to progress this far, because of the chance that it will be an enabling technology.

---

[6] 3D Technology Laboratories is in a sense a poster child for the Committee's analysis of *Government-Industry Partnerships for the Development of New Technologies*. The 3D technology is technically complex, results from research at a major U.S. university, and has a long lead time. At the same time, it also has multiple potential applications across a wide range of agency missions, from space exploration and defense to health care. Reflecting this diversity of applications, and the management's steep learning curve, 3D Technology Laboratories has made use of a surprising range of government programs to support new technologies, cited in the text. Relatively few companies do this, either because the technology would not qualify or because the management is unaware of the opportunities for federal funding.

"Without government support," she concluded, "new technology-based products cannot be developed. If this one lives for decades past what we are putting into it, then America as a country and our economy and the industries that can use this new type of visualization tool will benefit. My goal is to take it to the point where it can survive on its own." She is beginning to look for nongovernment sources of funding for the next stages of development.

DISCUSSANT

Jim Turner
House Science Committee

Mr. Turner agreed with other participants that the conditions at Ames amounted to a unique opportunity to develop partnerships, and that the geographical, technological, and other assets of Ames gave this lab an excellent chance of succeeding in its objectives.

He also offered several notes of caution. First, he reminded the audience that Congress is highly critical toward programs that bring any suggestion of "corporate welfare" or government giveaways—even for programs that are essentially self-financing. He suggested that Ames planners take special care in how they interact with private firms and that they pay special attention to how their plans might be perceived in Washington.

He added that it appears that Congress intends to create a new requirement that Phase 2 SBIR awards include a commercial plan, including steps in marketing and selling technology, and said that those companies who work with Ames would be well prepared for these new requirements.

He commented on Dr. Downing's unsatisfactory experiences with university partnerships, suggesting that while the Bayh-Dole Act had generally "proved itself" in the context of federal laboratories by transferring the rights of discoveries to the inventor, the allocation of IP rights at universities is still evolving. He proposed that she might find a more satisfactory partnership with a national lab, such as Ames.

In regard to the In-Q-Tel program, Mr. Turner said that an SBIR program for the CIA might have some advantages over a venture capital fund. He said that while the In-Q-Tel program may indeed "hit it big" with a useful and money-making technology, it would be highly visible money vulnerable to appropriation by Congress. He suggested that a more orthodox program to seek out existing technologies and issue small contracts might be a practical way to financing start-ups with less political risk.

Importantly, he urged the Ames planners not to use up their "very precious resource" of land too quickly with many small programs. A good strategy, he said, is to reserve enough land to accommodate changes in their strategic vision as the years go by.

Finally, Mr. Turner praised the involvement of UC Santa Cruz and Carnegie Mellon, and also urged Ames not to rule out relationships with Stanford and Berkeley, "two of the best computer schools in the nation, right in your back yard."

QUESTIONS & COMMENTS

Dr. Penhoet extended Mr. Turner's comment about the danger of too many objectives, adding that it is rare to be able to meet multiple objectives under the same program. A participant, David Audretsch, offered the good-humored objection noting that Berkeley, where Dr. Penhoet serves as a dean, has multiple mandates itself. Dr. Penhoet replied that Berkeley also has a 150-year history of managing programs, and even so, it has had programs with multiple goals that do not thrive. He also observed that as a businessman he had seen many failures in the use of venture capital for multiple objectives. "You can either try to make money on venture capital," he said, "or you can try to use it as a window on technology, but you are unlikely to do both well." He suggested that this is a constraint to keep in mind.

Dr. Ballhaus of Lockheed Martin agreed with Dr. Behrens that the Ames plan may resemble corporate venturing more closely than traditional venturing. In corporate venturing, he said, the projects that work are those you do not micromanage. When Lockheed sets up a partnership, he said, it takes only a minority position, late in development, because "larger corporations that are managing things strategically are usually mismatched with respect to startups that are moving in an entrepreneurial fashion." He mentioned some of the triumphs of Xerox PARC in the form of its "offspring," including 3Com, Adobe, and Bay Networks. All of them avoided a strategic relationship with Xerox because they were moving swiftly to markets that were no longer interesting to the larger corporation.

Dr. Wessner agreed with a comment by Jim Turner about "virtual" partnerships conducted at a distance. Collaborations with off-site researchers would reduce crowding and increase the national reach of NASA, he said. He also cautioned against too much focus on equity investments.

In conclusion, Jim Turner praised the effort at Ames as an innovative use of the space program's resources. Turner also reiterated his word of caution, advising that Ames take special care to avoid the perception of "corporate favors" while at the same time profiting from the synergies and management experience that the private sector can bring. Charles Wessner suggested that the flexibility of the Space Act, and its legitimacy, should be kept in mind as the project goes forward. Seed capital, from SBIR awards, and other arrangements already permitted under the Space Act might accomplish most of Ames' objectives without raising as many policy questions as would equity investments. The technological and perhaps political risk associated with equity investments or venture activities should be kept in mind. Success rates for investments, even when made by outstanding venture capital firms, may not be high enough to meet Washington's admittedly ill-defined expectations. Nonetheless, while keeping in mind these cautionary comments, Dr. Wessner suggested that it is only fair to observe that this ambitious initiative does address needs central to the NASA mission and provide a means of meeting educational needs which are equally central to the continued development of the region.

Concluding Remarks

Henry McDonald
Ames Research Center

Dr. McDonald thanked the presenters and discussants, and summarized the intent of the Ames strategic plan as enhancing "this center's ability to contribute to the fulfillment of NASA's mission." As a major part of that, he said, Ames will involve itself with people who are likely to help do that, including universities and industrial organizations. "We'll also get involved in the educational process as a natural fallout from this collaboration, and this will also train our next generation workforce. So we will connect ourselves rather directly to the mission of the Agency and judge ourselves on that basis."

He concluded by addressing the issue of affordable housing, which had been raised by several participants. He pointed out that the area had been crowded and expensive since before he immigrated to the United States from Scotland many years ago, "yet we manage to recruit some of the best scientists in the country because of the stimulating intellectual opportunities we have to offer them." Some of these scientists eventually leave for tenure at a university, he said, but Ames gets some of their best years. Current plans to use housing on site, and to develop plans to extend that housing, can help alleviate the problem and allow for the expansion of on-site programs and the inclusion of graduate students, postdocs, and summer programs for faculty.

In closing his remarks, and the symposium, Dr. McDonald again thanked the participants, noting that their expert but informal dialogue had sharpened the formulation of Ames' objectives in exactly the manner they had hoped when they asked the Academies' STEP Board to review their plans and objectives.

V
RESEARCH PAPERS

Science and Technology Parks at the Millennium: Concept, History, and Metrics

A Background Paper for Planners of the Ames S&T Park

Michael I. Luger
University of North Carolina at Chapel Hill

INTRODUCTION

As part of NASA's expanding strategy of leveraging federal resources with private sector activity and commercial technology, the NASA Ames Research Center is developing a science and technology (S&T) park at its 2,000-acre facility located in California's Silicon Valley. This paper provides an overview of S&T and related park developments around the world, as a way to provide NASA with a broader context for its planning activities. The paper briefly profiles the growth of the S&T park movement over the past 50 years. It then shows the diversity in park designs and concepts. The third section that follows describes four trends in park development that mark the early 21st century. The paper concludes with some comments about the use of parks as an economic development strategy. In particular: how do we know whether a park should be built, and how do we measure its success?

DEVELOPMENT OF SCIENCE PARKS IN THE LATE 20TH CENTURY

Since the Stanford Research Park was built in the early 1950s, many more such developments have been opened, both in the U.S. and abroad. Depending on how one defines "park," there are many hundreds in existence, many more have been closed and many others are still in the planning stage. Today, there are 295 members of the Association of University-related Research Parks (a U.S.-based organization), several hundred members of the International Association of Science Parks (IASP), and dozens of members of several country-based

> **Types of "Parks"**
>
> Research parks
> – Cater to R&D operations
> – Examples: Research Triangle Park, Stanford Research Park
>
> Science/technology parks
> – Focus on application of science and engineering to the development of new products and processes with commercial potential
> – Examples: Centennial Campus (North Carolina State University); University of Utah Research Park
>
> High-tech industrial (or agricultural) parks
> – Occupants engage primarily in production of relatively high value-added goods
> – Many parks in Asia
>
> Warehouse/distribution parks
> – Big boxes. But may incorporate high-tech elements (e.g., advanced logistics)
> – Includes "Global Transparks" built in Kinston, North Carolina, and in Thailand, on sites of decommissioned airfields.
>
> Office/headquarters parks
> – Sales functions, administrative activities; regional presence
>
> Eco-industrial parks
> – Input-output linkages among tenants optimized to minimize accumulation/discharge of waste and pollution
> – Not really a "park" but a region
> – Best known example: Kalundborg, Denmark
>   • 75 miles east of Copenhagen on coast
>   • Began in 1970s spontaneously; members trying to reduce costs and meet regulatory requirements

park membership organizations. Parks have been built in almost every state, and in at least 60 countries around the world.

The physical characteristics of these developments vary, reflecting differences in the host country's or region's level of development, and in the parks' objectives, industrial focus, and type of ownership. There are "research parks," "science and technology parks," "high-tech industrial or agricultural parks," "warehouse/distribution parks," office/headquarters parks," and "eco-industrial parks." (See box above.) The common elements among these different varieties of development include the following (as per the IASP):

- the existence of operational links with universities, research centers and/or other institutions of higher education;

- their use to encourage the formation and growth of knowledge-based industries or high value-added tertiary firms, normally resident on site; and
- the presence of a steady management team actively engaged in fostering the transfer of technology and business skills to tenant organizations.

Research/Technology Centers and Technopoli

In the literature and common practice at least two additional terms are used that are not types of parks, but are related to park development: research/technology centers and technopoli. *Research/technology centers* are physical facilities in which science and technology-related activities take place, including R&D, meetings, skill training, testing, and tele-conferencing, for instance. Research/technology centers are commonly used as anchors within parks: for example, the biotechnology and microelectronics/information technology centers built by the state within Research Triangle Park in North Carolina; centers for biotechnology, materials science, information technology, and microelectronics built by the government in the National S&T Development Agency park near Bangkok; and a training center for IT workers to be built as part of a new Palestinian initiative on the border of the West Bank and Israel.[1]

Technopoli are regions developed around several interrelated "knowledge" elements, including, but not limited to, science parks, research/technology centers, and universities. Technopoli require special planning, including infrastructure development, housing, and transportation, to make sure the elements work together. Prominent examples include Tsukuba Science City in Japan and Taedock Science Town in Korea. The Chinese (PRC) government is working on a plan to develop its largest metropolitan region—Chongqing—into a technopolis.[2]

DIFFERENT TYPES OF PARKS FOR DIFFERENT PURPOSES

While these parks share common elements, they differ in terms of

- objectives;
- size and physical layout;
- ownership and management;
- typical activities and occupants;
- links to universities and technology bases;
- incentives; and
- infrastructure, facilities and services.

---

[1] See TSG (The Services Group, Inc.) *A Feasibility Study for the Khadouri Technology Development Center*. Final Report to the U.S. Agency for International Development. Arlingon, VA: TSG, Inc., 1999.

[2] See Michael Luger, Deog Song Oh, and David Gibson, Editors, *Technopolis as Regional Development Policy*. World Technopolis Association. 1998.

The discussion below reviews these differing aspects of science and technology parks, and outlines costs and benefits to the host country or region and to the individual company locating in such parks.

Objectives

The design, services, and functions of a technology park are first a reflection of its basic purpose. Many countries fail to recognize the fundamental diversity of science and technology parks, and tend to view these projects as specialized industrial parks. But the purpose and forms of science and technology parks vary greatly. Common objectives of science and technology parks are to

- promote research and development in leading-edge technologies;
- serve as a "growth pole" strategy for the development of regions;
- encourage entrepreneurship and business development in technology areas; and
- generate exports and create high-tech jobs.

While none of these objectives are mutually exclusive, successful parks have generally had a clear focus and a limited set of objectives.

- **Research and development**. In some cases, parks are conceived as long-term instruments to transform economic bases from typically more traditional sectors to higher tech. Job growth in these instances must be measured over a longer period of time as new technologies are developed or different types of businesses are induced to locate in the region. A prime example is Research Triangle Park in North Carolina. The electronics, pharmaceutical, and telecommunications clusters now located there developed slowly over a forty-year period and gradually helped transform the central part of North Carolina from an economy based on agriculture and low-wage manufacturing to one based on high-tech R&D. Other examples are found in most advanced economies, including the science parks in Finland, Sweden, United Kingdom, South Korea, Japan, Singapore, and Taiwan.

- **Growth poles**. Other parks have been developed as so-called technopoles or growth poles. Parks have served as the cornerstone of growth pole strategy – as a way to move population from dominant cities—in Japan (in Tsukuba Science City and Kyoto), Korea (Taedok Science Town in Taejon), and Taiwan (Hsinchu Science City). In those cases, park development was coordinated with other investment strategies, for infrastructure, higher education and research, and housing. Other prominent examples include the Sophia Antipolis technopole in France and the Medeira Technopole in Portugal.

- **Incubation**. Another explicit objective of science and technology parks is to serve as an incubator to promote start-ups and business development in defined technology areas. While many parks—such as the Singapore Science Park's Innovation Centre—house incubator facilities on-site, a few parks are incubators themselves. A prime example of this is the Tefen Park north of Haifa in Israel that serves as an incubator for export-oriented technology companies.

- **Export generation**. Another category of science and technology parks aim to generate exports in international trade services and products. A leading example are the twelve Software Technology Parks in India that currently account for 70 percent of India's total software and IT services exports of US$4 billion. Other examples are the 80 science and technology parks in China, and the Agean Free Zone Technopark in Turkey.

Size and Physical Layout

Parks range in size from one large building in an urban setting—for example, the University City Science Park in Philadelphia, Pennsylvania, and several facilities in Germany—to several thousand hectares, such as the 8,000 hectare Sophia Esterel Science Park in France.

One common (if not universal) feature of science and technology parks is their physical attractiveness. Park developers believe that good design and natural amenities are necessary to develop a conducive work environment for knowledge-based industries. As a result, many parks are developed as beautiful campuses with office park facilities. A leading example is the Hsinchu Science-Based Industrial Park in Taiwan, which was deliberately developed to resemble facilities in Silicon Valley in order to attract diaspora Taiwanese engineers working in California.

Ownership and Management

Science and technology parks are owned by universities (University of Utah and Stanford Research Parks in the U.S.), government agencies (the National Science and Technology Development Agency Research Park in Thailand—NSTDA), by private companies (Kyoto Science Park), and by consortia of different public and private stakeholders.

The objectives of the parks reflect their ownership. University-owned parks tend to focus on university-originated technology and on building industry-university linkages. However, universities also see parks as potential sources of real estate revenue (Centennial Campus at North Carolina State University; Cambridge Research Park, U.K.). Parks sponsored by government agencies are typically part of regional or national development efforts. An increasing number of parks are privately developed and owned. Leading investors in these projects include U.S., French, British, Singaporean, Thai, and South African groups, most of which have a property development background.

Management of parks also varies, but the industry trend is toward professional management services and away from the "do-it-yourself" approach. Some universities and government agencies do continue to operate their own parks. Research Triangle Park, for example, is operated by a not-for-profit foundation that reports to an ownership team comprised of the region's universities and the state government. However, even under these management structures, outsourcing of professional services is becoming common.

Typical Activities and Occupants

Parks also differ in terms of their sectoral focus and industry orientation. Many parks tend to specialize in a few technology and industry areas, serving as "centers of excellence," promoting innovation in a particular area. Examples include the following:

- Singapore Science Park, Singapore—information technology and telecommunications;
- Hsinchu Science-Based Industrial Park, Taiwan—computers, peripherals, integrated circuits;
- Bangalore Software Technology Park, India—software and IT services;
- Taedok Science Town, South Korea—memory chips, aerospace;
- Software Technology Park, Brazil—software engineering;
- University City Science Center, U.S.A.—engineering, biomedicine, materials;
- Helsinki Science Park, Finland—biotechnology, food industry; and
- National Science and Technology Development Agency Science Park, Thailand—biotechnology, metals and material technology, electronics, and computer technology.

Government-run science and technology parks oriented to basic science and R&D typically host government labs. Examples include the NSTDA park in Thailand, the national science labs in the U.S. (Sandia, Los Alamos, and others), and Taedok Science Town in South Korea. Other science and technology parks resemble typical office or business parks, accommodating regional and international headquarters companies. Leading examples include Stanford Research Park, Cambridge Research Park, and Dublin Science and Technology Park in Ireland.

Links to Universities and Technology Bases

Most successful science parks had a meaningful connection with an institution of higher education. As noted, some parks have been developed by universities as sites for university-related activity (e.g., Stanford Research Park, U.S.A.; Cambridge Research Park, U.K.; Parque Industrial de la Universidad de Guadalajara, Mexico; The Australian Technology Park). Others have forged relationships with nearby universities (Amsterdam Science Park; Sophia Antipolis, France; NOVUM Research Park, Sweden; Patras Science Park, Greece; Tecnopolis Csata Novus Ortis, Italy). Parks developing in regions without institutions of higher learning have created them as part of the park's amenities to tenants. That approach, clearly, is costly. Increasingly, parks are connecting to universities and colleges electronically, making immediate proximity less important.

The second type of focus is around technology sectors, usually capitalizing on existing strengths in the regional industrial base and the local universities. Larger parks may have several foci (Research Triangle Park with electronics, pharmaceuticals and biotech, and telecommunications; NSTDA Park in Thailand around biotech, electronics, and materials science). But many parks focus more narrowly, and even use the focus in their name as a marketing ploy (Audubon Biomedical Science and Technology Park; Harry Hines Medical Research Park; Environmental Technology Center Neopoli Oy, Finland; Agro-Business Park, Denmark; Infopark, Budapest; and Kalundborg Eco-industrial Park, Denmark).

Incentives

Different parks provide, through their sponsoring entity, a wide variety of incentives for businesses. Those incentives tend to be largest when the park is part of the national or state government's economic development program. Israel's central government, for example, provides businesses moving to Tefen (and other designated locations) a benefit of 24 percent of their investment in building and equipment grants, or a ten-year income tax holiday. However, these types of incentives are usually available to all qualifying high-tech investments, whether or not physically located within a technology park.

However, a few countries have either adapted existing incentives (usually within free zone schemes) or developed new packages specifically for enterprises located within science and technology parks. For example, the software-oriented parks in India (such as Bangalore) have done well, in part, because of the favorable tax treatment accorded those businesses locating there. The favorable treatment extends to foreign capital and has been responsible for an inflow of investment.[3]

---

[3] See TSG (The Services Group, Inc.), *A Feasibility Study for the Khadouri Technology Development Center, op.cit.*

Infrastructure, Facilities, and Services

Unlike most general industrial parks, science and technology parks emphasize purpose-built infrastructure and facilities, tailored to meet the requirements of target industries and activities. The range of facilities typically found include the following:

- research and testing labs—funded by government and major private corporations;
- business and technology incubators—operated by specialized subsidiary companies or independent operators on a commercial basis, providing a full range of business, marketing, legal, financial, and technical support services for start-up firms;
- high-tech office buildings with research units—usually pre-fabricated "intelligent" office buildings, for use on a multi-tenant basis with shared business support facilities and local area networking connections;
- standard factory buildings suitable for a variety of manufacturing and warehousing activities;
- residential, commercial, and recreational areas for employees and managers;
- exhibition areas, convention centers, and libraries;
- training and consultancy center—typically attached to an incubator or testing facility;
- dedicated, high-speed telecommunications facilities, offering high-speed (1.5 mbps) 7/24 lines at international prices, as well as value-added network services; and
- centralized support services including dedicated power, hazardous waste collection and disposal, as well as a range of business services at reduced rates (e.g., management training, technical assistance, procurement assistance, liaison with nearby universities and businesses, regulatory approvals, etc.).

The overall objective is to create a conducive work environment that enhances worker productivity and promotes technological collaboration and innovation among a cluster of inter-related companies.

BENEFITS OF SCIENCE AND TECHNOLOGY PARKS

Depending on the type of park, industrial focus, extent of government funding, additionality of investment—the magnitude of economic benefits from science and technology parks varies significantly. The value of a technology park is also different for each potential beneficiary—the host country or region, private companies, or participating universities.

Host Country. From the perspective of host countries and regions, science and technology parks provide a number of potential benefits. The most important of these include the following:

- *technological development*—parks offer the potential for industrial upgrading, research and technological innovation in high-tech areas;
- *cluster development*—parks can create self-sustaining industrial clusters in core technologies, and lead to the development of technology corridors in a wider area;
- *job generation*—parks are an efficient means of creating high value-added jobs in leading technologies;
- *business efficiency*—parks can enhance the operating competitive, image and investment environment of a region; and
- *university-industry linkages*—parks can offer a concrete mechanism for collaboration between universities and industries, and a focal point for technology transfer.

Assessing Economic Impact

The economic impact of science and technology parks is difficult to estimate, given the large variations on types of parks worldwide. The science and technology parks that exist account for a significant part of high-tech manufacturing and services, especially in developing countries. The software technology parks in India, for example, account for 70 percent of the export earnings of the software sector overall. Selected examples of these types of projects are profiled in the following table:

Economic Impact of Science and Technology Parks—Some Examples

Technology Park	Size	Established	Firms	Jobs
Singapore Science Park, Singapore	30 hectares	1980	226	7,000
Rennes Atalante Science & Technology Park, France	70 hectares	1978	250	8,000
Hsinchu Science-Based Industrial Park, Taiwan	580 hectares	1980	272	72,623
University City Science Center, Philadelphia, U.S.A.	7 hectares	1963	140	7,000
Kyoto Research Park, Japan	8.5 hectares	1988	80	2,400
National Technological Park, Ireland	260 hectares	1991	90	3,500
Technopark Kerala, India	73 hectares	1994	35	2,000
Surrey Research Park, U.K.	28.5 hectares	1974	76	2,000

Source: TSG, 1999

The overall economic impact of a technology park depends on a number of factors. In conducting a benefit-cost analysis of the welfare consequences of park development, one must account for the opportunity cost of government subsidies in those cases where the public sector provides key infrastructure and services. In those cases, a park may appear to be thriving, but may have a low public internal rate of return. Strictly private parks, therefore, tend to have a higher incidence of "failure" since they are subjected to the rigors of the market.

The net economic impact of a park also depends on the extent to which investments and employment are truly additional, and would not have taken place anyway in the absence of a park. The net impact of a project is also reduced if most investments are simple relocations of companies already operating elsewhere in the country, though management willingness to relocate may suggest an appreciation of the region's positive externalities over and above any relocation incentives. Backward supply linkages of high-tech industries also tend to be low, until a critical mass of local, non-park industries develops over time, which of course is one of the long-term objectives of park planners.

Individual company. Benefits from a science and technology park location for an individual company vary again depending on the scale and type of investment. For small-scale and start-up investments, for example, the total package of facilities, support services, and technical and financial resources available through a park are a major attraction. In general, location within a technology park—rather than outside—provides firms with a number of benefits:

- access to a nucleus of technology resources and specialized services in one area;
- scope for collaboration with other technology companies and suppliers;
- access to better-quality, purpose-built infrastructure and facilities and competitive prices;
- reduction in costs through the provision of shared services and facilities;
- superior quality of life and amenities;
- acquisition of highly specialized knowledge, often tacit, through access to a pool of workers, technicians and scientists, with partner universities and institutes; and
- access to a competitive package of investment incentives.

The major drawback of some technology park locations—particularly as projects expand and mature over time—is the possibility of increased labor turnover. Employees can more easily jump from one company to another, given the proximity of similar companies in one area. But these are less prevalent in science and technology parks compared to general industrial parks.

FOUR RECENT TRENDS IN S&T PARK DEVELOPMENT

Parks being developed at the dawn of the new millennium exhibit four new trends:

- increasingly, they are built around one or several key (core) national sectors;
- the intensified competition for R&D means parks are tying into existing and emerging clusters;
- universities affiliated with parks are applying the "green door" concept more frequently (that refers to a door between a scientist's academic lab and commercial lab, located close for easy access); and
- virtual parks are now being developed.

Core Sectors

Increasingly, parks are developed around one or several core sectors, for several reasons. First, as the number of parks around the world has increased, competition for high-tech businesses to come to each park has intensified. Recruiters can be more effective if they focus on a few industries. That way, they get to know the needs of the businesses better and the key decision makers, via networking. That focus also is a marketing tool for the park, allowing the recruiter to hail the location as the "place to be" for photonics or biotech or other technologies. Related to that, park developers can build (or induce the location of) support services and amenities appropriate for the specific target industries. This focus on key industries is related to the use of cluster analysis, discussed next.

RTP, the Research Triangle Park in North Carolina, illustrates this approach. From its very beginnings, it focused on microelectronics and pharmaceuticals. The North Carolina state government built microelectronics and biotechnology research centers to help attract those industries. As the park matured, so did its targeting strategy, focusing more on IT and communications, and less on microelectronics. The Royal Thai government took a similar approach in developing its large park in Rangsit: they built four research centers in the technology areas that were the focus of development: biotech, materials science, electronics, and infomatics. A new park, being developed by USAID in the West Bank of Palestine, is focusing on software companies, blending the supply of software engineers in Palestine with the supply of capital in Israel. A key element of that plan is a software center that will provide training and space for start-up companies. And in the plan for the Ames S&T Park, intelligent systems, high performance computing, and aviation operations systems are among the targeted sectors.

Cluster Analysis

With intensified competition for R&D, parks are tying into existing and emerging "clusters" more and more. A cluster is a group of firms and related institutions where the competitiveness of any firm is dependent on the competitiveness of other members of the cluster.[4] Clusters can be formed among businesses related through input-output linkages, among firms that use the same types of labor and support services, or among businesses that share the same market.

One implication of this recognition of clusters is that a "park" may be spread over several locations in a region (or more broadly in the case of a "virtual" park, described below). One park location may house high-end R&D and another may be the site of the companies' prototype production, with both functions being connected via fiber, microwave, and/or satellite links.

One interesting application of this concept are the Global Transparks being developed in North Carolina and on the east coast of Thailand. These "parks" are being built along air strips (both of which had previously had military use). The Asian site will ship component parts manufactured in Asia using relatively cheap labor. Those parts (modules) would then be flown to North Carolina to be assembled, largely along a robotocized assembly line, and then delivered to market in the U.S. and Europe. Orders would come to the North Carolina plant electronically. That order would trigger a chain of supply orders that would be completed with very little need for inventory. Speed to market would be achieved by the electronic nature of the process, but also because time is saved moving goods from Asia to the U.S., across the international dateline. An order shipped on Monday from Thailand would arrive in North Carolina the same day. If the final product were then going to the Central, Mountain, or Pacific time zones, additional hours would be gained. A park is planned for Frankfort, Germany, to save time further. Then, components would be shipped straight to Germany from Thailand, and then to European markets.

A New University Connection

As indicated in the IASP definition on the first page of this paper, a university connection is a common feature of all science parks. Luger and Goldstein show, as well, that U.S. parks tend to be university-owned, university-operated, or somehow affiliated with universities.[5] That university connection has trans-

---

[4] See Edward Feser, *High Tech Clusters in North Carolina*. Report Prepared for the North Carolina Board of Science and Technology. Chapel Hill, NC: Office of Economic Development, 2000. See also Michael Porter, *The Competitive Advantage of Nations*. New York: Free Press, 1990.

[5] See Michael Luger and Harvey Goldstein, *Technology in the Garden: Research Parks and Regional Economic Development*. Chapel Hill: University of North Carolina Press, 1991.

lated into joint research, easy access to jobs for students, faculty consulting opportunities, adjunct appointments for industry scientists and engineers, and some joint facilities.

The new university connection is what Thomas Meyer (Associate Director of Los Alamos National Labs) calls the "green door" concept: the location of a park so close to academic researchers' labs that all they need to do is go through the green door and work on the commercial side of their science. He proposed such a park at the University of North Carolina. One has been built at neighboring North Carolina State University.

North Carolina State University's Centennial Campus

- University has strong engineering, agriculture, and textile programs
- Was given a 500-hectare plot of land next to campus
- Built an Engineering Graduate Research Center and moved their School of Textiles there
- Built labs and incubator space for university departments, and for rental and purchase by private companies
  - companies had to prove they were working with faculty (they are called "partners")
- Very strong performance: mix of larger companies' R&D (Lucent Technologies) and start-ups
  - 60 partners, 900 scientists and engineers matched with 900 faculty and staff engaged in work

---

As this inset shows, Centennial Campus has been a tremendous success in terms of the demand for space by "partners." Those are companies that buy or lease space (including in new incubators) on the NCSU campus, with the explicit intent of working with university researchers. This has increased the production of intellectual property at the university considerably.

Virtual Parks

With the advent of increasingly affordable high-speed communications via fiber optics, microwave, and satellite, businesses do not have to have propinquity to be connected, and therefore can be part of a virtual park. Those parks may be owned and managed by the same group, who helps the scattered businesses maintain their connection, and still provides common services, but now, over the web.

Many companies have maintained connections with its own branches via teleconferencing facilities. Now, unrelated companies are growing their connections too, for contract negotiations, shared training, and joint research with each other and university partners.

In the United States, the National Science Foundation recently funded a program called "collaboratories," which was intended to bring together via a teleconference the best minds from around the world, applied to critical problems.

In Asia, a prominent developer is building a virtual park by investing in high-speed telecommunications hardware to service the site. He expects to appeal to multinational corporations who value real-time connectivity with their branches, headquarters, and other businesses.

DEFINING AND MEASURING SUCCESS

Economic development policy makers face two critical questions:

- How do we know whether a park *should* be built?
- How do we judge *whether it has been* successful?

The first of these is the *ex ante* policy analysis question; the second is the *ex post* policy evaluation question.

Ex ante Analysis

In a classic policy analysis, we compare interventions to each other relative to their performance against pre-selected criteria. These criteria include short-run rate of return, which is important on the real estate side, as well as longer-term efficiency.

Market analysis can help determine whether there is sufficient demand to make the model that is being developed work. That analysis can help design the type of park, or can indicate whether a park would be viable at all. The importance of this is underscored by the statistics: one out of every four parks that started-up between 1954 and 1990 failed altogether as a real estate project. Half of the surviving parks had to change their focus to remain viable.[6]

Market analysis extends to an assessment of whether the proposed site has the right "fundamentals" available: well-priced land, access to customers electronically or directly, adequate labor supply with appropriate skills, other traditional and knowledge infrastructure, and a good quality of life.

The efficiency question amounts to asking how to get the biggest bang for the (public sector) buck: is a park development the best use of resources to achieve a given set of objectives? To answer that question we need a clear specification of objectives. We also need to sort out whether benefits are private or public (for example, spillover benefits to universities).

---

[6] See Michael Luger and Harvey Goldstein, *Technology in the Garden*, op.cit.

Ex ante analysis also includes an assessment of the political and institutional environment, specifically, whether there is support for the project in terms of complementary policy and leadership to get past regulatory hurdles and any potential opposition. Are the key stakeholders lined up in support? Is the concept consistent with strategic planning for the region and state? And is the concept based on a realistic reading of the emerging economy?

Ex post Evaluation

Program evaluation is becoming more common in government. At the federal level, the Government Performance and Results Act of 1993 required government agencies to develop strategic plans and then file annual reports marking progress to the articulated goals. Congress and state legislatures have instituted sunset provisions in some legislation, requiring performance reviews before funds are renewed. Benchmarking is becoming more popular at the state level, and in the area of S&T policy, indicators are becoming more and more prevalent. (NSF has just released a major RFP toward that end.)[7]

From a practical standpoint, these efforts to look systematically at programs' progress toward goals have several benefits for policy makers. Good results strengthen the claim on resources. Questionable results provide an opportunity for planners to fine tune or change the program. That is important since one of the "critical success factors" in S&T park development is adaptability—the revision of objective, change in focus, and alteration of programs in response to changing market conditions and new opportunities.

Ex post evaluation has different uses when done short-term and long-term. The short-term assessment—done a year or two after implementation—is cruder, but still useful for fine tuning programs. Long-term assessment is useful to judge the net social benefits (via cost-benefit analysis). It requires more data (over more years). It also becomes more difficult because the benefits are both direct and indirect. Efficiency assessment also requires the evaluator to separate gross from net (or induced) effects.

Application to S&T Parks

A major *ex post* evaluation of S&T parks in the United States concluded that successful parks tend to have a number of common attributes:

- strong leadership;
- visionary planning;

---

[7] For a summary of these developments, see Catherine Renault, Leslie Stewart, and Michael Luger, *Economic Development Evaluation and Monitoring System for North Carolina*, Report prepared for the North Carolina Department of Commerce, July, 31, 2000.

- deep pockets and patience;
- good timing;
- appropriate services; and
- meaningful relationships with universities.[8]

The authors were emphatic, however, that these do not constitute a menu for success. While they tend to be correlated with success, they are neither necessary nor sufficient for positive results. Local context is critical. There are nuances and subtleties that arise in every individual case.

IMPLICATIONS FOR NASA AMES

Some lessons can be gleaned for NASA Ames from this overview of S&T park development. First, the traditional real estate criteria for success are abundantly present: the very availability of developable land in the heart of Silicon Valley, proximate to major high-tech corporations and world-class universities, bodes well for the marketability of the development. Second, the scientists and engineers at NASA Ames are engaged in high-powered research that would seem to have considerable commercial potential (assuming the science is not classified). Third, the NASA research budget is substantial and, therefore, attractive to university and industry researchers who seek contracts and joint research opportunities. The planners of the Ames S&T Park project need to evaluate for themselves whether there is sufficiently strong leadership, visionary planning, deep pockets patience, appropriate services, and meaningful relationships with universities.

In terms of the design of the project, to be consistent with 21st century trends, planners of the project and university partners should provide support to maximize the commercial payoff through new product development and spin-off companies. That includes meaningful connections among university, government, and private sector researchers (including implementation of the green door concept) and the establishment of virtual connections with distant researchers and organizations. The location of facilities on-site by Carnegie-Mellon is an example of that. NASA Ames also might consider a focus on a subset of its technology areas, choosing technology foci that correspond with existing and emerging strengths in the industry base of the region.

Finally, in order to use resources efficiently and to ensure continued governmental (NASA and Congressional) support it will be important to incorporate the *ex ante* and *ex post* evaluation procedures into the planning and implementation of any park.

---

[8] See Michael Luger and Harvey Goldstein, *Technology in the Garden, op.cit.*

REFERENCES

Feser, Edward. 2000. *High Tech Clusters in North Carolina*. Report prepared for the North Carolina Board of Science and Technology. Chapel Hill, NC: Office of Economic Development. See also www.rri.wvu.edu/WebBook/Bergman-Feser/contents.htm.

Luger, Michael and Harvey Goldstein. 1991. *Technology in the Garden: Research Parks and Regional Economic Development*. Chapel Hill: University of North Carolina Press.

Luger, Michael, Deog Song Oh, and David Gibson, editors. 1998. *Technopolis as Regional Development Policy*. World Technopolis Association.

Porter, Michael. 1990. *The Competitive Advantage of Nations*. New York: Free Press.

Renault, Catherine, Leslie Stewart, and Michael Luger. 2000. *Economic Development Evaluation and Monitoring System for North Carolina*. Report prepared for the North Carolina Department of Commerce. July 31.

TSG (The Services Group, Inc.). 1999. *A Feasibility Study for the Khadouri Technology Development Center*. Final Report to the U.S. Agency for International Development. Arlington, VA: TSG, Inc.

The Prospects for a Technology Park at Ames: A New Economy Model for Industry-Government Partnership?

David B. Audretsch
Indiana University

INTRODUCTION

There are now hundreds of science and technology (S&T) parks in the United States, and thousands in the world. Some four decades ago they were an adventurous undertaking – poorly understood, often conceived with little insight as to what realistic goals should be or how progress might be evaluated. In the subsequent years, however, much has been learned from first-hand experience and assessment of S&T parks. The motives, rationale, ingredients for success and indicators for evaluation and monitoring are not only more transparent but also better understood.[1]

This long experience with S&T parks would make it seem that evaluating the prospects for the proposed NASA Ames Research Park should be a straightforward undertaking. After all, the large number of S&T parks would seem to provide the appropriate benchmarks to enable a confident assessment of the Ames Research Park proposal.

This is not the case. Ames is different. As is explained and documented in the second section, the traditional S&T park has a mandate to transfer technology that has been produced within the knowledge source of the park outwards

---

[1] Michael Luger and H. Goldstein, *Technology in the Garden: Research Parks and Regional Economic Development*, Chapel Hill: The University of North Carolina Press, 1991, pp. 174-184. More recently, see National Research Council, *Industry-Laboratory Partnerships: A Review of the Sandia Science and Technology Park Initiative*, Charles W. Wessner, ed., Washington, D.C.: National Academy Press, 1999.

for commercialization by private industry within the region. The primary goal of traditional S&T parks is to provide an engine of growth for the region via outward technology transfer. Thus, the benchmarks and measurements evaluating the impact of the traditional S&T park is typically in terms of regional economic development—the (quality) jobs created, new firms generated, branch plants and corporate headquarters attracted, and increases in the regional growth rates induced.[2] A traditional S&T park that cannot document any changes in these indicators over a long period would surely be classified as a failure.

But Ames is different. Ames is different because the primary goal of the research park is not to foster regional economic development. In fact, Ames is located in Silicon Valley, which is the most technologically developed region in the world. The rise in both incomes and employment in Silicon Valley has been unrivaled in the world. Between 1992 and 1996, employment increased by 15 percent. At the same time, wages rose to a level that is 50 percent greater than the national average.[3] If any region in the United States does not need to worry about economic development, it would be Silicon Valley. The challenge for Silicon Valley is managing and sustaining its unprecedented economic growth.

This does not mean that the Ames Research Park proposal is superfluous. Rather, as explained and documented in the third section, the goals of the Ames Research Park are markedly different than those found in traditional S&T parks. The fundamental goal of the Ames Research Park is to enable NASA to achieve its mission by providing economical access to technological capabilities external to NASA. There are two main ways that economic technological access will be achieved. The first is through the inward transfer of technology developed outside of NASA. The second is through the joint development of new technology by NASA in conjunction with external partners in private industry and the universities.

While the traditional industry-government-university partnerships involving S&T parks involves an outward flow of knowledge from the government or university research facility to private industry, in the Ames Research Park the flow is reversed. Thus, the traditional S&T model is about getting a higher return from government investment in research by facilitating commercialization in the private sector. The Ames Research Park is also about getting a higher return from government investment in research. However, the difference is that in the case of Ames, the government investment in research is used to leverage access to research capabilities and competencies in the university and industry sectors. Commercialization still plays an important role, but it is radically different. Rather than serving as the mechanism generating regional growth, it instead provides

---

[2] Luger and Goldstein, *Technology in the Garden*, *op. cit.*, pp. 14-33.
[3] David B. Audretsch and Roy Thurik, *Innovation, Industry Evolution, and Employment*, Cambridge: Cambridge University Press, 1999, p. 5.

the carrot to entice private industry participation to assist in achieving national goals.

Similarly, excellence in university research capabilities is the starting point for commercialization in the traditional S&T model. In the case of Ames, participating in the NASA research park is the incentive to universities for upgrading research excellence.

As is made clear in the fourth section, this different mandate for the Ames research park requires a different perspective on benchmarking and measuring its impact. A major difference revolves around focusing on the increased and more economic ability of Ames to meet its mission as a result of the Research Park. Additional benefits will also be accrued to universities in terms of increases in educational programs, and to private industry in terms of new and more economical innovative activity in key technologies.

In the final section of the paper conclusions are provided. In particular, the NASA Ames Research Park may represent a new model for industry-government partnerships. As knowledge plays an increasing role in the New Economy, this model may become more prevalent than the more traditional industry-government partnership.

THE TRADITIONAL SCIENCE & TECHNOLOGY PARK MODEL

The Traditional S&T Park

Science and technology (S&T) parks are a phenomenon of the post-war era. The first S&T park may have been the Stanford Industrial Park, which was opened in the early 1950s. While a number of other parks were created in the subsequent years, the majority of S&T parks were founded in the 1980s and 1990s. There are currently hundreds of S&T parks in existence in the United States. In addition, there are now S&T parks in over 60 other countries, including Western Europe, Japan, and Australia.[4]

Because of the complex and ambiguous missions, defining S&T parks has proven elusive. Still, in their essence, S&T parks are intended to serve as a *seedbed* or catalyst for the development of a cluster of innovative- and technology-oriented business enterprises in a region or state. The Association of University-Related Research Parks (AURP) provides a definition, which includes the following components:

- existing or planned land and buildings designed primarily for private and public research and development facilities, high-technology and science-based companies, and support services;

---

[4] Luger and Goldstein, *Technology in the Garden, op. cit.* pp. 14-33. See also Michael Luger, "Science and Technology Parks: Concepts, History and Metrics," in this volume.

- a contractual and/or formal ownership or co-operational relationship with one or more universities or other institutions of higher education, and science research;
- the promotion of new ventures and economic development; and
- a role in aiding the transfer of technology and business skills to the industry tenants.

The S&T park may be a not-for-profit or for-profit entity owned wholly or partially by a university or a university-related entity. Alternatively, the park or incubator may be owned by a non-university entity but have a contractual or other formal relationship with a university, including joint or cooperative ventures between a privately developed research park and a university.

There are five distinct types of traditional parks:

- *Innovation Centers*: Within or alongside a university campus, these provide small units for firms growing out of research or expertise within the university. They are usually housed in existing buildings. These are the research environments that transform basic inventions into commercially viable innovations.
- *Science & Research Parks*: These are developments designed for growing or established firms in research and development that can be associated with university research laboratories and ancillary amenities. They have workshop, laboratory, and office functions. Science and research parks are typically joint ventures between the private sector and a tertiary educational institution, although they do not need to be sponsored or funded by these organizations.
- *Technology Parks*: These comprise establishments that undertake a high proportion of applied research, possibly but not essentially involving a university. To be successful they require high-quality housing in the immediate vicinity and university and research institutions within a thirty-mile radius. The character of the physical and social environment is an important prerequisite in order to attract scientific and professional staff. These developments are almost invariably constructed with a low building density in attractively landscaped settings.
- *Commercial/Business Parks*: These involve high-quality, low-density environments with accommodation intended for commercial firms requiring a prestigious image and a high-caliber workforce. They do not require a link with an academic institution but need to be essentially attractive to a mixture of manufacturing, sales, support, and professional service functions.
- *Upgraded Industrial Parks*: There are a great number of straightforward industrial park developments that have aspired to the research park image and are presented and marketed as such. While they have little or no direct

connection with knowledge-based research activities, their quality of design and appearance has benefited as a consequence of the visual standards spilling over from the *bone fide* high-technology sector.

Park Objectives

While the definitions of traditional S&T parks remains elusive, the objectives are remarkably singular and focused:

1. the promotion of technology transfer from the laboratories to the development of tenant companies;
2. the stimulation of new technology-based startups; and
3. the attraction of mobile R&D projects of large companies.

S&T parks vary considerably in their organizational, managerial, and locational characteristics, with parks now in place in 42 states. They are located in urban areas of all sizes, ranging from the largest metropolitan areas to small cities hundreds of miles from the nearest metropolitan area. Some parks are situated in old, rehabilitated factory or warehouse buildings in dense parts of central cities, while others are laid out along winding roads in low-density, green, campus-like, suburban environments. Around one-quarter of the parks are units of public or private universities. State or municipal governments account for another 16 percent. Slightly more than one-fifth of the parks are non-profit corporations or foundations. Fifteen percent of existing parks are owned by for-profit corporations, while the remaining 21 percent are joint public-private ventures.

The size of research parks, measured in aggregate employment, ranges from no employees to 32,000. About one-third of the parks have no employment at all. While the mean workforce of research parks is about 1,700 employees, the skewed size distribution results in the large majority of research parks having a workforce of fewer than twenty employees.

Finally, S&T parks differ in the managerial strategies and policies that managers adopt. These differences, in turn, reflect differences in the particular objectives of the park. For example, many parks target the R&D branch plants of multilocational corporations, while others focus on generating start-ups with local entrepreneurs and nurturing small, innovative-oriented business start-ups. The nature of the physical facilities—for example, the overall land use density in the park and the existence of multi-tenant buildings and incubators—and the types of services provided by the park management often reflect the strategy pursued.

Benefits from Traditional S&T Parks

The widely accepted premise underlying the traditional research park strategy is to promote the economic development of the region. Thus, the locus of benefits is typically at the regional level. The task of the S&T park is to generate a transfer of technology from the laboratory to private industry. The benefits to the region are accrued in terms of jobs created, new firms created, and growth.

For regions faced with a high cluster of older, declining manufacturing industries, S&T parks have been viewed as the vehicle for industrial restructuring. For other regions where the economy has been performing well, S&T parks represent a long-term investment strategy. In both cases, the R&D-led regional economic development strategy, when successful, almost always leads to more than just employment growth and new business formation. It also brings with it concomitant changes in the employment mix, wage and salary structure, political culture, and spatial patterns of economic development.

Most studies evaluating S&T parks are anecdotal, describing park characteristics (inputs), rather than outcomes (performance). When the success or failure of a park has been systematically and quantitatively analyzed, it has been as a real estate venture alone.

Potential Primary Impacts of Science and Technology Parks on Regional Economic Development

Type of Impact	Immediate Source of Impact	Mechanism	Comments
Location of new R&D activity	Park enterprises, university, other R&D activity, milieu	Localization Economies	Growth will depend on amount of R&D in the region, strength of region's universities in tech-related areas, and/or presence of government research labs.
R&D firm Spin-offs	R&D enterprises in park; scientific faculty brought to region	Localization Economies	The rate of spin-off activity varies by enterprise ownership, type of R&D activity in and out of park, and university regulation of faculty entrepreneurship.
Location of Manufacturing Activity	R&D enterprises in park and induced R&D activity	Backward Linkages	Material factor inputs are a small fraction of total R&D costs; leakage from region is typically high but varies by type of enterprise and degree of vertical integration.
Business Services Location	R&D enterprises in park; induced R&D manufacturing and other functions	Backward Linkages	Depends on enterprise ownership, types of R&D firms, and any induced manufacturing firms.

Potential Primary Impacts of Science and Technology Parks on Regional Economic Development—Continued

Type of Impact	Immediate Source of Impact	Mechanism	Comments
Intrafirm Manufacturing Location	R&D enterprises in park; induced R&D and manufacturing in region	Forward Linkages	Depends on importance and frequency of face-to-face contact between R&D and manufacturing functions within the firm and on the corporate organization of R&D.
Location of other Intrafirm Functions	R&D enterprises in park; induced R&D and manufacturing in region	Forward Linkages	Depends on enterprise ownership, type of R&D and manufacturing activities, proximity of R&D and HQ functions, supply of skilled labor. Large metro most likely to attract HQ and sales functions.
Retail and Consumer Services Growth	New households from induced migration to region's labor force	Earnings Multiplier	Magnitude depends on total amount of induced growth (first 6 types of impacts) and the new workforce's pay level. Minimum leakage from the region.
Generalized new Business Development	All sources listed above	Urbanization Economics	The larger the region, the higher the magnitude. Amenities and the quality of public management may be important.
Increased productivity of existing firms	R&D activity by park enterprises and in region's university	Technology Transfer	Depends on match between R&D and technology needs of region's industries. Innovation adoption rates vary between existing and new firms and by effectiveness of marking services to region's firms.
Loss of business (industrial gentrification)	Park enterprises and induced enterprises with high pay	Age wage Roll-out	Depends on magnitude of wage/salary differences between existing and new firms, and ability to transfer labor skills.

The benefits of S&T parks have been categorized by Luger and Goldstein[5] as consisting of primary economic growth impacts, distributive dimensions of primary impacts, and secondary (economic structure) impacts. Primary impacts include the effect of changes in the magnitude of economic activity—for example, the number of businesses and jobs, personal income, and value added. The primary impacts, which shape regional economics, include induced growth in R&D activity, manufacturing activity, business services and headquarters functions, retail and consumer services, productivity of firms in the regions, and the loss of existing businesses. These primary impacts have some relevant distributional dimensions—spatial, occupational, and socioeconomic. Secondary, or derivative, impacts are those that are induced from the primary changes but also result in changes in the economic structure of the region. The distributive dimensions of the primary impacts include the impact on the skill and education requirements of the occupational categories, and the enterprise structure (single plant, locally owned versus multilocational firm). The secondary (economic structure impacts) are measured in terms of changes in a region's economic stability, enterprise/ownership mix, productivity, product mix (by position in the product cycle), wage structure, in/out-migration patterns, labor force participation rate, structural unemployment rate, poverty and unemployment rates, level of income equality, land and housing prices, and labor-management relations.

Success Factors in S&T Parks

Why are some S&T parks more successful than others? Studies[6] have identified seven key factors that shape the success of S&T parks. These factors are university involvement, the presence of high-tech talent, project funding, physical infrastructure, entrepreneurial culture, amenities, and leadership.

University Involvement

Although the necessary conditions for S&T park success are far from unambiguous, the presence of a large research university appears to be quite important. The presence and involvement of a major research university is a characteristic common among virtually every successful S&T park.

---

[5] Luger and Goldstein, *Technology in the Garden, op. cit.*, pp. 34-48. See also discussion of evaluating parks in Luger's paper in this volume.

[6] Rolf Sternberg, "The Impact of Innovation Centres on Small Technology-Based Firms," *Small Business Economics*, 2(2): 105-118, 1990; and "Technology Policies and the Growth of Regions," *Small Business Economics*, 8(2): 75-86, 1996.

One reason why university involvement is the key to S&T park success is that they possess or have access to the critical knowledge resources, such as scientific and medical equipment, trained students in search of professional work experience, and highly trained research faculty. Universities are also connected to large public and private funding sources (often with a long-term focus), providing powerful collaborative opportunities. Resource sharing between firms and universities also serves as a major benefit of collaboration.

The most important reason for involving a university comes from the advantages of having academic and private researchers working in close proximity to one another. When academics and firms collaborate on projects, share common facilities or interact with each other, the likelihood for knowledge transfer and the creation of new knowledge increases. Universities are the world's largest loci of knowledge. S&T parks that collaborate with universities located on or very near their park campus generate a culture of open exchange, interaction, and innovation. This culture enables park participants to share existing knowledge, which can be used to increase a collective stock of knowledge in their professional communities much more rapidly.

Presence of High-Tech Talent

In order for an S&T park to be successful, a critical mass of knowledge workers needs to be in the region. As Luger[7] points out, if a S&T park is analyzed purely from a real estate perspective, it must attract a minimum number of companies in order to survive. Such companies come from high-tech talent.

Project Funding

Successful S&T parks share a common characteristic of long-term funding. While there are a wide variety of sources of funding for S&T parks, many of these sources provide funds only for the first couple of years, with the expectation that the S&T park will be financially viable and sustainable. However, this is not the case of the traditional S&T park. Thus, long-term funding makes a large difference in enabling S&T parks to develop.

There are a number of sources providing funding and/or resources for technology-based start-ups. For example, the Small Business Innovation Research (SBIR) Program provides grants to businesses for innovative research in areas where there is a high potential for commercialization.[8]

---

[7] Luger and Goldstein, *Technology in the Garden*, *op.cit.*, pp. 14-33.

[8] For an overview of this $1.2 billion program, see National Research Council, *The Small Business Innovation Research Program: Challenges and Opportunities*, Charles W. Wessner, ed., Washington, D.C.: National Academy Press, 1999.

Physical Infrastructure

A well-developed transportation network, with emphasis on the highway system and proximity to a major international airport, enhances accessibility of an S&T park. This feature is of particular importance for ensuring the mobility of resources—supplies, products, and people. The physical infrastructure is a complementary asset in attracting conferences, visiting scientists, researchers, and businesses. In an increasingly digitalized economy, communication networks are also essential to successful S&P parks. The presence of high-speed fiber optic communication lines facilitates video conferencing and rapid transfer of data.

Entrepreneurial Culture

Since the traditional S&T parks have focused on technology transfer from the S&T park to the private sector, in order to serve as an engine of regional economic development, the existence of an entrepreneurial culture has played an important role. The ability and willingness of individuals and teams of individuals to commercialize some of the knowledge in the S&T park by starting a new firm serves as a key vehicle for this knowledge transfer. Some analysts suggest that a number of S&T parks that have not been successful lacked such an entrepreneurial culture.[9]

Amenities

A high quality of natural and social environments contributes to the overall quality of life. Although the desirable combination of natural amenities may differ according to personal preferences, most individuals place a high value on clean air, water, and the natural surroundings, and tend to find places plagued by high levels of pollution and crime undesirable. A high-quality social environment typically includes the presence of good quality residential areas, elementary and secondary schools, hospitals, and access to public facilities, such as museums, entertainment, and other forms of recreation. Additional social amenities include established, proven universities, tertiary education establishments, and research institutions. A high quality of amenities is a prerequisite for attracting knowledge workers to the region.

Dedicated Leadership

Dedicated leadership champions the enterprise and characterizes successful parks. For example, the success of the technology incubator in Austin, Texas,

---

[9] Amy Glasmeier, "Factors Governing the Development of High-Tech Industry Agglomerations: A Tale of Three Cities," *Regional Studies*, 22(4): 287-301, 1987.

can be attributed to significant leadership from Henry Cisneros, who was then the mayor of San Antonio, and George Kozmetsky, the dean of the Business School. The governor in 1983, Mark White also served as a champion of transforming Austin to a new high-technology region. Recent analysis has also documented the important role that leadership played in the formation of Research Triangle Park in North Carolina.[10] In particular, the champion function of dedicated leadership is important for the following:

- land management, including sales and leasing;
- financial management and income collection;
- organizing the maintenance of the grounds and shared facilities;
- gaining representation in local and/or state policy formulation;
- strategically attracting the start up of new firms;
- strategically attracting the location of existing firms;
- coordination of private and public actors; and
- provision of legal standing and policies toward legal issues such as intellectual property rights.

AMES: A NEW MODEL FOR INDUSTRY-GOVERNMENT PARTNERSHIP?

The NASA Mission

The future mission of NASA involves challenges such as a single stage to orbit launch vehicles, upgrades to the shuttle and International Space Station operations, an earth science sensing fleet, planetary sample return, advanced aircraft concepts, human exploration, next generation astronomy, and near-sun measurements. Accomplishing this mission will require a new set of capabilities for NASA.

To meet its exceptional mandate, NASA must develop means to overcome the barriers of time, distance, and extreme environments. This will require NASA to develop future systems that are *autonomous*, that is, systems that will have the capacity to think for themselves. These systems will require the capability for evaluating uncertain situations and undertaking actions in uncertain environments. This will require the ability to create information and knowledge from data and to generate greater productivity with fewer people. Technology will be substituted for human decision-making.

In addition, future systems will need to be *resilient*, in that they are highly durable and damage tolerant. Rather than relying upon external assistance for

---

[10] Albert N. Link, *A Generosity of Spirit: The Early History of the Research Triangle Park*, Durham, NC: Duke University Press, 1995, pp. 25-36.

repair, they will have the capacity to perform self-diagnosis and repair. Equipment durability will increase even as the external conditions become increasingly harsh. These systems will be *evolutionary*, in that they have the capacity to adapt the form and function to meet changing demands and overcome unanticipated problems, as well as to grow and expand to exploit new opportunities. These systems will also be *self-sufficient* in that they require minimal on-board resources. They will be cut off from the "base camp" of Earth and need to "live off the land" of their own environment.

Overcoming the barriers of time, distance, and extreme environments will also require future systems to be developed that are *highly distributed*, in that they provide broad, continuous presence and coverage, as well as interactive networks to achieve maximum capability at the most efficient use of resources. Thus, these systems will require *ultra-efficiency* in their use of mass, power, and volume, enabling travel about the Earth and universe to be rapid, safe, and cost-efficient.

Thus, the future challenge of NASA is to develop systems that can provide these capabilities in order to overcome the barriers of time, distance, and extreme environments. These systems will be based on new and revolutionary technologies, which combine nanotechnology, biotechnology and information technology.

Nanotechnology involves the creation of functional materials, devices, and systems at the nanometer scale and enables the exploitation of novel properties—physical, chemical, and biological—at that scale. The technology will make it possible to develop sensors, actuators, devices, and lightweight structural materials at an unprecedented small scale. These products are the key to developing a new generation of aerospace transport vehicles and "thinking" spacecraft and systems.

Nanotechnology provides a basis for miniaturizing biochemical analytic laboratories, such as nano-devices and sensors that enable the detection and characterization at the quantum limit of single photons, cosmic particles, and molecules. This will facilitate the detection of subtle signatures of life and provide a characterization of deep space objects.

In addition, nano-structured materials will enable orders-of-magnitude enhancement in structural materials properties and integrated structural, computational and sensor functionalities. This will make it possible to develop microsatellites for planetary and small-body exploration and huge apertures to characterize extrasolar planets, facilitating the study of phenomena under extreme conditions, such as black holes.

Nano-structural engineering will enable adaptivity and reconfigurability at the molecular level. It will also facilitate the merging of software and hardware for biometric system responses to changes in internal or external conditions. Nano-structural engineering may make it possible to develop self-repairing spacecraft, self-reconfiguring space systems to optimize mission return, and space system lifetimes of decades to centuries for interstellar exploration.

The projected future budgets of NASA do not provide adequate resources to meet the stated agency mission. The proposed NASA Ames Research Park would provide one mechanism for leveraging the limited resources of NASA with the private sector and major universities to contribute to the NASA mission.

A Unique Goal

Thus, the proposed NASA Ames Research Park has a very different goal than that of traditional S&T parks. While traditional S&T parks are oriented toward transferring technology from the knowledge source to the external regional community, the goal of the Ames Research Park is to provide the internal knowledge source—NASA—economically efficient access to knowledge and capabilities either found in the external community or which a strategic partnership could develop more efficiently and economically.

Multiple Means

This goal would be accomplished by the following:

- establishing strategic partnerships with major companies and universities in key research areas such as astrobiology, information technology, nanotechnology, and biotechnology;
- exploiting existing and developing new facilities for such collaborations;
- creating new opportunities for NASA education programs;
- contributing resources to spread the fixed costs of operations; and
- enhancing workforce capabilities through
  - joint appointments and internship programs;
  - access to graduate students, "post-docs" and future employees; and
  - on-site workforce continuing education.

The Role of Universities

Universities will provide one leg of the strategic triad upon which the Ames Research Park will be based. In order for the NASA Ames Research Park to succeed, mechanisms have to be developed to facilitate the interaction of Ames scientists with the university research community. Such mechanisms are provided in the Ames Research Park design.

The UC Partnership

The University of California at Santa Cruz has been selected as the lead for the overall University of California System as the strategic partner with Ames. Under this strategic partnership, the NASA Research Park will be designated as the preferred Silicon Valley site for regional research and education. This partnership should provide NASA with a vehicle for gaining access to the resources of the University of California system as well as a basis for collaborative research.

The resources that NASA will be able to access at the University of California system are formidable, and include ten campuses with three Department of Energy national laboratories. With an annual budget of $13.6 billion, the University of California has an annual research budget of $2.0 billion. Of the 7,000 faculty, 40 have been awarded Nobel Prizes and 300 are National Academy of Sciences Fellows.

One of the important assets that the University of California system will bring to the NASA Ames Research Park is a strong link to commercial biotechnology firms. In fact, one-third of U.S. biotechnology firms are founded within 35 miles of a UC campus. In California, home of the largest number of biotechnology firms, one-quarter of the companies were founded by University of California scientists, including Amgen, Chiron, and Genetech. In addition, 85 percent of the biotechnology firms in California employ alumni of the University of California system with graduate degrees.

Providing access to NASA of the strong link between the University of California system and the private biotechnology sector should yield benefits in NASA's mission to develop a Center for Star Formation, an Astrobiology Institute, Remote Sensing, Data Visualization, Mars Missions, and Space Biology.

The strategic partnership also envisages joint tenured appointments between Ames and the University of California. In addition, graduate students will work collaboratively between the university and NASA. Provisions are also made for the formation of joint research teams and for the creation of new and unique collaborative research facilities.

NASA will participate in the development of the CASC teacher institute along with workforce development for high-tech employment. To extend the partnership beyond the University of California system, a consortium will be formed involving San Jose State University and Foothill-DeAnza Community College.

There will be benefits from this strategic partnership for the University of California System, but particularly for the Santa Cruz campus. The partnership will create a new model for science education, which brings together the strengths of government, industry, and the university. Included in this new educational model are novel and innovative outreach programs focusing on the digital divide, and joint doctorates and research with San Jose State University and NASA. This new model should strengthen UC Santa Cruz and support its leadership's effort to make one of the most prominent research universities in the world. In particular, the collaborative research agenda between UC Santa Cruz and NASA will result in the UC Santa Cruz being the lead research university for the Carl Sagan Astrobiology Laboratory, and enhanced research and teaching capabilities in the fields of biotechnology, information technology, nanotechnology, planetary sciences, K-12 and teacher education, and the digital society.

The Carnegie Mellon Partnership

A second initiative providing a framework institutionalizing interaction between Ames' scientists and the university research community is provided by a strategic partnership with Carnegie Mellon University. This partnership will have an initial focus on robotics and high reliability computing, two of the traditional strengths of Carnegie Mellon. Ames will provide Carnegie Mellon students with internships. In addition, the partnership will form the basis for consortia with Silicon Valley companies.

The partnership will provide NASA with access to the research resources and scientists on the Pittsburgh campus as well as provide a gateway to Silicon Valley for Carnegie Mellon scientists and graduates. It is anticipated that NASA and Carnegie Mellon will develop some unique educational programs to meet the needs of the partnership.

From its experience in partnerships with the Robotics Institute (with Westinghouse) and the Software Engineering Institute, Carnegie Mellon has learned that education complements research and is an essential component of technology transfer. The proposed partnership will involve collaborative research with NASA and other universities as well as companies located in Silicon Valley. This partnership should yield benefits for NASA's space mission, because Carnegie Mellon has extensive experience and expertise in developing robotic systems. This competence will be the basis for joint research on reliability, autonomy, robot team coordination, robotic work systems, and robotic exploration and discovery. This research is expected to yield valuable applications for life seeking in extreme environments and planetary global exploration.

Industry Participation

Private industry is a key player in the NASA Ames Research Park model. Potential industry partners, such as Lockheed Martin Corporation, argue that there is a critical mass of shared objectives to make the partnership successful. In particular, the complementary research assets in key technologies such as astrobiology, information technology, nanotechnology, biotechnology, life and microgravity sciences, and aeronautical and space technology provide potential gains to both industry partners as well as NASA.

Joint research should be promoted through the creation of unique facilities and laboratories for research collaboration, as well as workforce enhancement, such as the joint appointment of scientists, internship programs, graduate students, and doctoral students. In addition, continuing education programs sponsored jointly by NASA and private industry should enhance the workforce of both partners.

The Research Initiative Fund

To support these goals, Lockheed Martin proposes the creation of a Research Initiative Fund, which would be held in an interest-bearing escrow account. The application of the Initiative Fund would be determined jointly by the three legs of the Ames Research Park triad—NASA, private industry, and the University of California at Santa Cruz. The funds would be used for grants for academic fellowships, funding for NASA research programs, and the development of new mechanisms to promote research.

The facilities at Ames, including the buildings, would be used in a manner that utilizes the complementary assets between NASA and private industry. In terms of Lockheed Martin, this would involve facilities dedicated to information technology, including computer hardware, software, internet, electronics, broadcasting, and telecommunications, as well as astrobiology, aviation, aerospace, biotechnology, and nanotechnology.

Managing the Tripartite Model

The success of the NASA Ames Research Center depends not just on the conception of the model but also on how it is managed. The issue of Center management revolves around developing mechanisms and tools for NASA to access the resources of strategic partners and to focus them on meeting goals consistent with NASA's missions. Only through developing such instruments can the Ames Research Park reach its full potential.

The Entrepreneurial Center

One key instrument for park management proposed by the Commercial Technology Office of Ames is to create an *Entrepreneurial Center*. This Center will expand the pool of NASA technology resources through focused partnerships with industry. These partnerships will be selected to accelerate the fulfillment of NASA mission requirements. In particular, initiatives undertaken by The Entrepreneurial Center will seek to resolve common technology problems, accelerate spin-offs of NASA technology to the private sector, and expand opportunities for NASA incubators. The objectives of The Entrepreneurial Center are to create focused and dynamic commercial partnerships. For its part, NASA is to provide laboratory space, scientific expertise and experience, access to NASA technologies, and a long-term research focus. In return, the strategic partners will provide industry expertise, a greater awareness of potential commercial applications, better access to venture capital and new-venture finance, and business experience, as well as an overall industry presence at Ames. These complementary assets are expected to result in benefits for both partners. In particular, these partnerships are expected to give NASA the capabilities to address problems on an industry timetable, rather than a government timetable.

To facilitate The Entrepreneurial Center, the Commercial Technology Office can also rely on its existing tools, which include the following:

- technology assessment;
- marketing;
- intellectual protection and licensing;
- agreement development;
- regional and national industry networks;
- management of the Ames Small Business Innovation Research Program; and
- business incubation.

Management responsibilities for The Entrepreneurial Center include identifying potential technologies appropriate for collaboration, preparing a finite project plan, and implementing the project upon approval. The Commercial Technology Office will be charged with approving a project plan, establishing and approving access to NASA labs, and establishing and approving access to NASA researchers.

Potential Barriers

It is anticipated that as a result of collaborations with industry and universities, NASA will be able to leverage its resources to effectively double its investment. However, in order to accrue the benefits of strategic partnerships with industry and universities, a number of barriers and hurdles must be overcome.

Intellectual Property: One set of barriers involves issues surrounding the competing needs for intellectual property rights for each partner. Since the joint product of research collaboration is intellectual property, each partner has a vested interest in holding the rights to that intellectual property. Unless new models can be developed from sharing and/or allocating the intellectual property accruing from the joint research, industry and universities will be hesitant to join in such partnerships.

Decision-Making: A second set of barriers involves bureaucratic processes. The pace of government is considerably slower than industry. Government processes typically require massive paperwork in decision-making processes and for the approval of programs and initiatives. In addition, government activity is generally placed under a barrage of rules and regulations, making it considerably more rigid than private industry. Partners from the industry and university sectors are not likely to be patient and tolerant of such bureaucratic barriers, which could ultimately subvert the partnership. One important issue determining the success of the NASA Ames Research Park is the selection of strategic partners. Selecting the right partners will ensure that synergies are created that generate benefits for all parties. Selection of inappropriate partners will result in wasteful investments yielding few benefits. Ames' criterion for selecting strategic partners are based on four aspects:

- the strategic partnership results in an activity supporting the mission of NASA under the Space Act;
- the strategic partnership involves the appropriate use of Federal property;
- the strategic partnership is consistent with site environmental constraints; and
- the strategic partnership is consistent with local community needs and priorities.

This broad selection framework provides appropriate standards for the selection of appropriate strategic partners.

The NASA Enterprise Fund

Another instrument for park management is the *NASA Enterprise Fund*. The business concept underlying the NASA Enterprise Fund is the establishment of a technology investment fund that is market driven and has a return on investment criterion. NASA can be included as a limited investment partner, drawing upon a portfolio of a $900 million annual technology program and a $100 million annual Small Business Innovation Research (SBIR) program. The SBIR projects span 18 major technology areas and cover around 95 projects per year. This will enable venture partnerships with NASA based on technologies not normally seen by the investment community. In addition, NASA will be able to participate in partnerships in an effective collaborative manner that is not limited by the traditional constraints and rigidities associated with government. These new ventures will be targeted in investment areas identified as critical to the NASA mission – information technology, nano-technology, MEMS, compact sensors, and biotechnology.

The NASA Enterprise Fund will provide access to new technologies being developed by private industry as well as accelerated technology development.

For the Fund to succeed it needs to provide

- means of bringing high-technology investment opportunities not normally available to or recognized by the venture finance community;
- means for entrepreneurial firms to grow based on profits on technologies developed via venture investment shared by NASA; and
- opportunities for technical risk reduction in new ventures as a result of the technical participation by NASA.

Two things in particular would doom the NASA Enterprise Fund: first, if it is seen as a direct competitor in the venture business community; second, if the government rules, regulations, and general bureaucracy associated with normal operations become extended to the Fund. To succeed, the Fund must avoid becoming a direct competitor with the venture investment community, as well as the imposition of government rules, regulations, and constraints.

The NASA Enterprise Fund should prove to be a successful management tool for leveraging NASA's technological assets and gaining access to resources in the private and university sectors. This is because the Fund is based on bringing together the complementary research and technology assets of NASA with those in the private sector. NASA will gain by the creation of a profit center for innovative technologies focusing on NASA mission technology initiatives. The Enterprise Fund should provide NASA with the opportunity to accelerate technology and acquire technology from future commercial markets at costs that are substantially lower than if NASA had developed those technologies alone.

The investment community should view the Enterprise Fund as an opportunity for profitable investments based on leveraging NASA technologies, and to reduce some of the risks associated with research and innovation in technologies where NASA has an expertise.

MONITORING AND MEASURING THE TECHNOLOGY PARK IMPACT

Because the goals and mission of the Ames Research Park are markedly different from that of traditional S&T parks, monitoring and measuring the impact of Ames must reflect this difference. As explained in the second section, the approach to monitoring and measuring the impact of traditional S&T parks has been to focus on the flow of knowledge from the park to the external region with a particular emphasis on commercialization, job creation, and growth.[11]

However, the logic of Ames is radically different. The main mission of Ames is to facilitate the attainment of NASA's mission. Thus, the flow of knowledge is much more from external partners into NASA. At the same time, much

---

[11] See the discussion in Luger's paper in this volume.

of the new knowledge is anticipated to emanate from the interaction among all three partners—NASA, private industry, and universities. In the case of Ames it would be inappropriate to monitor and measure the impact in terms of the usual criteria applied to judge the impact of traditional S&T parks, such as new firms created, new jobs generated, establishments and corporations locating in the region, and change in regional growth.

Rather, the impact of the Ames Research Park must be measured in terms of the benefits to the three major participants compared to the counterfactual situation if no such research park existed. However, the economic attainment of NASA's mission must carry the greatest weight in measuring and monitoring the impact of the Ames Research Park:

- **Economic Attainment of NASA's Mission**
  This involves measuring the extent to which attainment of NASA's targeted technologies are attained at a cost below that which NASA would have incurred if it had developed the technologies by itself. In addition, it involves measuring and placing a dollar value on the accelerated time development of such technologies. There are a number of intermediate measures that are important indicators of the impact that the Research Park is having in facilitating the Ames mission. These include
  - changes in the numbers and impact of patents filed jointly with Research Park partners;
  - changes in the numbers of published articles and citations with Research Park partners;
  - changes in different types of interactions between NASA and the external scientific community; and
  - changes in the quality of the Ames and NASA workforce that is recruited and sustained.

- **Improvements to Educational Institutions**
  A different set of benefits is relevant for the impact on universities. These benefits focus on the impact that the Ames Research Park has on education and on the participating (and non-participating) universities and other educational institutions. In particular, the education delivered is compared to the counterfactual benchmark of what would be delivered in the absence of the research park. Measures and benchmarks need to capture educational gains that otherwise would not have occurred. Intermediate measures of the impact on education include
  - the value of new programs and numbers of students enrolled, and graduates from the programs;
  - the quality of new faculty attracted as a result of the new programs; and
  - the changes in the output of the participating university departments and programs.

- **New and More Efficient Technologies Developed by Private Industry**
  There are also benefits accruing from the Research Park in terms of new commercial products and technologies that otherwise would not have come into existence, or would have come into existence at a greater cost over a larger timeframe. The appropriate metric would be the dollar value of costs incurred developing new technologies jointly with NASA and other Ames Research Park partners compared to the costs that would have been incurred in the absence of such partnerships. Such measures and monitoring will require the assessments of experts familiar with the technology and costs of research. Intermediate measures to indicate these types of gains include
  - joint patents between NASA and private industry;
  - joint publications in scientific journals by NASA and private industry;
  - changes in the workforce as a result of the partnership with Ames; and
  - measures of new-firm startups resulting from the Ames Research Park, such as the number of new startups, employment in firms started at the park or as a result of the park, numbers of IPOs, and value of external finance invested.

Hardest to measure may be initiatives that can be undertaken cooperatively with the involvement of the Ames Research Center and its partners which otherwise would not have been undertaken at all.

CONCLUSIONS

The traditional S&T parks were founded on the premise that a government and/or university institution had the competitive advantage in the production of knowledge over private industry. The goal of the S&T park was to provide an instrument for channeling that knowledge into commercialization opportunities for private industry. Through the flow of knowledge from the source within the park to commercial opportunities in the region, the traditional S&T park served as an engine for regional economic development.

The Ames Research Park is founded on the very different premise that private industry is no longer at a competitive disadvantage in the production of knowledge, but is at least an equal, if different, partner. In order for NASA to attain its mission, access to the knowledge resources in the private industry and university sectors is required. Thus, the flow of knowledge is no longer outward, with the aim of regional economic development, but rather inward and interactive, with the goal of enabling the government agency to achieve its goal by accessing the complementary knowledge assets in the industry and university sectors.

As private industry becomes increasingly based on knowledge in the New Economy, the Ames Research Park model for an interactive industry-government partnership is likely to become more prevalent than the one-way knowledge flows found in the more traditional model of industry-government partnerships.

REFERENCES

Audretsch, David B. and Roy Thurik. 1999. *Innovation, Industry Evolution, and Employment.* Cambridge: Cambridge University Press.

BankBoston Economics Department. 1997. *MIT: The Impact of Innovation.* Boston, MA: BankBoston Economics Department.

Feller, I. 1990. "Universities as Engines of R&D-Based Economic Growth: They Think They Can" *Research Policy.* 19(4): 335-348.

Glasmeier, A. 1987. "Factors Governing the Development of High-Tech Industry Agglomerations: A Tale of Three Cities." *Regional Studies.* 22(4): 287-301.

Glasmeier, A. 1990. *The Making of High Tech Regions.* Princeton: Princeton University Press.

Link, Albert N. 1995. *A Generosity of Spirit: The Early History of the Research Triangle Park.* Durham, NC: Duke University Press.

Luger, Michael. 1987. "The States and Industry Development: Program Mix and Policy Effectiveness." in J.M. Quigley (ed.). *Perspectives on Local Public Finance and Public Policy.* Greenwich, CT: JAI Press. pp. 29-64.

Luger, Michael. 2000. "Science and Technology Parks at the Millennium: Concept, History, and Metrics" in this volume.

Luger, Michael and H. Goldstein. 1991. *Technology in the Garden: Research Parks and Regional Economic Development.* Chapel Hill: The University of North Carolina Press.

National Research Council. 1999. *Industry-Laboratory Partnerships: A Review of the Sandia Science and Technology Park Initiative.* Charles W. Wessner, ed. Washington, DC: National Academy Press.

National Research Council. 1999. *The Small Business Innovation Research Program: Challenges and Opportunities.* Charles W. Wessner, ed. Washington, DC: National Academy Press.

Saxenian, A. 1994. *Regional Advantage: Culture and Competition in Silicon Valley and Route 128.* Cambridge, MA: Harvard University Press.

Sternberg, R. 1990. "The Impact of Innovation Centres on Small Technology-Based Firms." *Small Business Economics.* 2(2): 105-118.

Sternberg, Rolf. 1996. "Technology Policies and the Growth of Regions." *Small Business Economics.* 8(2): 75-86.

VI
ANNEX

Annex A

Ames White Paper on the Research Park

NASA has a bold new vision for the 21st century to partner with local communities, government, academia, private industry, and non-profit organizations. The goal is to establish a world-class, shared-use education and R&D campus featuring partnerships in astrobiology, information technology, aerospace, education, and commercialization.

> "Not from NASA alone, not from Silicon Valley industry alone, and not from world-class universities alone will tomorrow's *required innovations emerge.* This will come from all of us working together and making the most of the special attributes each of us brings to the table. That is what we will do at Ames."
>
> —Daniel S. Goldin, Administrator, NASA

This is NASA's vision for a bold new way of doing business at Ames. This vision includes goals for collaboration, business incubation, and education.

NASA Research Park is one component of the 2,000-acre Ames Research Center. The Research Park site was transferred to NASA in 1994 from the Navy as a result of the Base Realignment and Closure Act. With its prime Silicon Valley location, prominent architecture, and availability of land, NASA Research Park will be an ideal place where NASA, its collaborative partners, and the public can come together to expand human understanding of the origins of life on Earth, promote advances in aerospace and aviation technology, and understand advances in technology though public displays, interactive exhibits, lectures, and school programs.

NASA plans to create a unique community of research scientists, students, and educators with a shared mission to advance human knowledge of space, the Earth, and society. A lively and vibrant community will attract industry. To support this community, NASA, directly or through its collaborative partners, will offer support services and programs such as child care, housing, retail goods, business support services, meeting spaces, overnight accommodations, and recreational opportunities. In addition, NASA will provide critical public safety services and other services typically furnished by municipal government.

In partnership with academia and industry, NASA will promote entrepreneurship and innovation at NASA Ames Research Park. By taking advantage of its proximity to leading entrepreneurs and heads of innovative organizations, NASA and its partners can support the development of business incubators focused on the high-technology and bio-technology industries. Linkages can be formed with business education programs to provide forums, seminars, executive lecture series, and other venues to facilitate the exchange of information and experience to solve real-world business problems related to technology innovation, technology commercialization, and technology management.

NASA seeks partners who are compatible with NASA's mission at Ames Research Center, possess the financial capacity and experience to implement their proposed occupancy, and accept NASA's minimum business terms. Prospective collaborative partners will be evaluated based on established criteria. Primary factors of importance are the degree of collaboration in support of NASA's and Ames' mission, the degree of educational and learning programs supporting NASA's mission, and partnerships fostering business incubation and technology transfer.

The activities of Ames Research Center are governed by the National Aeronautics and Space Act of 1958 (42 U.S.C. § 2451 et set.) as well as other applicable laws. NASA has several available authorities under which partners can use and occupy buildings. Reuse of the buildings and new construction at NASA Research Park must be for purposes that are consistent with the National Aeronautics and Space Act. The Space Act has several provisions relating to agreements, including concession contracts, leases, cooperative agreements, and permits. NASA has authority to use a "Reimbursable Space Act Agreement" for certain building occupancy transactions involving non-federal entities. Reimbursable Space Act Agreements are based on cost recovery from on-site parties engaged in research relating to NASA's mission. NASA also utilizes the leasing authority granted to federal agencies under the National Historic Preservation Act of 1966 to enter into "historic leases" for specific buildings.

NASA's management and governance responsibilities include providing overall management of NASA facilities and NASA Research Park; ensuring compliance with the Space Act and all other applicable federal laws, regulations, and NASA policies; establishing programmatic guidelines and goals and communicating these to existing and prospective partners; monitoring adherence to the development plan and approving any modifications to or amendments of the plan; providing technical assistance, particularly in science education, research program development and program management to existing and prospective partners; identifying and managing improvements to facilities and infrastructure; adopting and applying design and construction guidelines for historic and non-historic properties; and monitoring rehabilitation and construction activities of on-site partners.

NASA's primary financial goal for the NASA Research Park is to leverage existing federal appropriations to support the maximum level of research and development, education programming, and learning opportunities possible. To accomplish this goal, NASA seeks to obtain cost reimbursement from NASA partners, charge appropriate rent for historic properties in NASA Research Park to ensure the integrity of the historic district, and generate new funds for collaborative scientific research from new construction within NASA Research Park.

A Master Plan for the physical lay-out and characteristics has been completed and a financial feasibility analysis has been performed on the concept. The concept includes retention and enhancement of the existing historic district, construction of a new Astrobiology Laboratory, development of the California Air and Space Center in Hangar 1, construction of a Computer History Museum, incorporation of the California Center for Business of the Future, establishment of one or more university campuses, protection and expansion of natural and specie habitat, utilization of a new light rail station, and total build-out of approximately three million square feet of office, laboratory, retail, and institutional space.

Annex B

Biographies of Contributors

David B. Audretsch

David B. Audretsch is the Ameritech Chair of Economic Development and Director of the Institute for Development Strategies at Indiana University. He is also a Research Fellow of the Centre for Economic Policy Research (London). He was at the Wissenschaftszentrum Berlin fuer Sozialforschung in Berlin, Germany, which is a government-funded, research think tank between 1984 and 1997. Between 1989 and 1991 he served as Acting Director of the Institute. In 1991 he became the Research Professor. Audretsch's research has focused on the links between entrepreneurship, government policy, innovation, economic development, and global competitiveness. He has consulted with the World Bank, National Academy of Sciences, U.S. State Department, United States Federal Trade Commission, General Accounting Office and International Trade Commission, as well as the United Nations, Commission of the European Union, the European Parliament, the OECD, as well as numerous private corporations, state governments, and a number of European Governments. He is a member of the Advisory Board to a number of international research and policy institutes, including the Zentrum fuer Europaeisch Wirtschaftsforschung (ZEW, Centre for Economic Research), Mannheim, Germany, the Hamburgisches Welt-Wirtschafts-Archiv (HWWA, Hamburg Institute of International Economics), and the American Institute for Contemporary German Studies (AICGS), Washington, D.C. His research has been published in over one hundred scholarly articles in the leading academic journals. He has published twenty-five books, including *Innovation and Industry Evolution*, with MIT Press. He is founder and editor of the premier journal on small business and economic development, *Small Business Economics: An International Journal*. He was awarded the 2001 International Award for Entrepreneurship and Small Business Research by the Swedish Foundation for Small Business Research.

Michael I. Luger

Michael Luger is a professor of public policy analysis, planning, and business at the University of North Carolina at Chapel Hill (UNC-CH). He is the founding director of the university's Office of Economic Development. He previously was the Carl H. Pegg professor of planning and chairman of the Curriculum in Public Policy Analysis at UNC-CH. He taught previously in the economics departments at Duke University and has been a visiting faculty member at the University of Maryland and Wirtschaftsuniversitat Wien.

Dr. Luger's research is on topics in economic development, science and technology policy, infrastructure finance, and urban and regional economics. His most recent book, published in December 2000 by Rutgers University Press, is *Red Tape and the Cost of New Residential Development*. He is the guest editor of a forthcoming issue of the *Journal of Comparative Policy Analysis*, on information technology and regional development. He has written widely about the role of science parks and universities in economic policy, including the book, *Technology in the Garden: Research Parks and Regional Economic Development*. He has served as a consultant and advisor on science parks and other S&T development strategies to the EU, UNIDO, OECD, the World Bank, USAID, the governments of Japan, Korea, Thailand, Palestine, China, and others.

Professor Luger received his Ph.D. in economics and an M.C.P. (planning) from the University of California, Berkeley, and an M.P.A. (public and international affairs) and A.B. (architecture and planning) from Princeton University.

Annex C

Participants List*

Duane Adams
Carnegie Mellon University

James Arnold
NASA Ames Research Center

David B. Audretsch
Indiana University

William Ballhaus
Lockheed Martin

Kathy Behrens
Robertson Stephens Investment Management

Jan Behrsin
University of California

Thomas Berndt
NASA Ames Research Center

William Berry
NASA Ames Research Center

Nancy Bingham
NASA Ames Research Center

Carolina Blake
NASA Ames Research Center

John Boyd
NASA Ames Research Center

Lewis Braxton
NASA Ames Research Center

Robert Caret
San Jose State University

McAlister Clabaugh
National Research Council

Jana Coleman
NASA Ames Research Center

---

* Speakers are *italicized*

Elizabeth Downing
3D Technology Laboratories

Maylene Duenas
NASA Ames Research Center

Leroy Fletcher
NASA Ames Research Center

William Foster
Invisible Studios

Lori Garver
NASA Headquarters

James Gill
University of California
   at Santa Cruz

M.R.C. Greenwood
University of California
   at Santa Cruz

Anthony Gross
NASA Ames Research Center

Susan Hackwood
California Council on Science
   and Technology

Warren Hall
NASA Ames Research Center

Robert Hansen
NASA Ames Research Center

Marla Harrison
NASA Ames Research Center

Diana Hoyt
NASA Headquarters

Cliff Imprescia
NASA Ames Research Center

Robin Kennedy
NASA Ames Research Center

George Kidwell
NASA Ames Research Center

Stephanie Langhoff
NASA Ames Research Center

Vic Lebacz
NASA Ames Research Center

Zoe Lofgren
U.S. House of Representatives

Gilman G. Louie
In-Q-Tel

Michael I. Luger
University of North Carolina

Patrick Mantey
University of California
   at Santa Cruz

Michael Marlaire
NASA Ames Research Center

Connie Martinez
University of California
   at Santa Cruz

Sally Maudlin
NASA Ames Research Center

Henry McDonald
NASA Ames Research Center

Burton McMurtry
Technology Venture Investors

ANNEX C

Meredith Michaels
University of California
   at Santa Cruz

James Morris
Carnegie Mellon University

Thomas Moyles
NASA Ames Research Center

Mark Myers
Xerox Corporation

Robert Norwood
NASA Headquarters

Edward Penhoet
University of California at Berkeley
   and Chiron Corporation

Robert Rosen
NASA Ames Research Center

Allison Rosenberg
University of California

Ken Souza
NASA Ames Research Center

James Turner
House Science Committee

Thomas Vani
University of California
   at Santa Cruz

Samuel Venneri (via
   videoconference)
NASA Headquarters

Mark Weiss
Xerox Corporation

Charles Wessner
National Research Council

Robert Wilson
University of Texas at Austin

Patrick Windham
Stanford University
   and Windham Consulting

Steve Zornetzer
NASA Ames Research Center

Annex D

Bibliography *

Alic, John. 1992. *Beyond Spinoff: Military and Commercial Technologies in a Changing World.* Boston: Harvard Business School Press.

Allen, David N., and David J. Hayward. 1990. "The Role of New Venture Formation/Entrepreneurship in Regional Economic Development." *Economic Development Quarterly* 4(1): 55-63.

Allesch, J., and H. Fieldler, eds. 1985. *Management of Science Parks and Innovation Centers.* Berlin.

Ambrose, Stephen. 2000. *Nothing Like It In the World: The Men Who Built the Transcontinental Railroad, 1863-1869.* New York: Simon & Schuster.

Association of University Related Research Parks. 1997. (www.aurrp.org).

Audretsch, David B., and Roy Thurik. 1999. *Innovation, Industry Evolution, and Employment.* Cambridge: Cambridge University Press.

BankBoston Economics Department. 1997. *MIT: The Impact of Innovation.* Boston: BankBoston Economics Department.

Barro, Robert J. 1990. *Macroeconomic Policy.* Cambridge, Massachusetts: Harvard University Press.

Berglund, Dan, and Chris Coburn. 1995. *Partnerships.* Columbus, Ohio: Batelle Press.

Berry, Brian J. L. 1973. *Growth Centers in the American Urban System.* Cambridge, MA: Ballinger.

Branscomb, Lewis. 1999. "The False Dichotomy: Scientific Creativity and Utility." *Issues in Science and Technology.* 16(1).

Branscomb, Lewis. 2000. *Taking Technical Risks: How Innovators, Managers, and Investors Manage Risks in High-Tech Innovations.* Cambridge, MA: MIT Press.

Branscomb, Lewis M., and James H. Keller, eds. 1998. *Investing in Innovation: Creating a Research and Innovation Policy.* Cambridge, MA: MIT Press.

Braun, Bradley M. 1992. "Science Parks as Economic Development Policy: A Case Study Approach." *Economic Development Quarterly* 6(2): 135-147.

Brown, Wayne S. 1987. "Locally-Grown High Technology Business Development: The Utah Experience," pp. 177-83. *Entrepreneurship and Technology: World Experiences and Policies.* W. Brown and R. Rothwell, eds. Harlow, Essex: Longmans.

---

* Includes selected references from the papers as well as a variety of references of general interest.

Castells, Manuel, and Peter Hall. 1994. *Technopoles of the World: The Making of Twenty-First-Century Industrial Complexes.* New York: Routledge.

Cohen, Linda, and Roger G. Noll, 1991. *The Technology Pork Barrel,* Washington, D.C.: The Brookings Institution.

Cox, R. N. 1985. "Lessons from 30 years of Science Parks in the U.S.A," pp. 7-25. *Science Parks and Innovation Centers: Their Economic and Social Impact.* J.M. Gibb, ed. Amsterdam: Elsevier Science Publications.

DeVol, Ross C., et al. 1999. *America's High-Tech Economy: Growth, Development, and Risks for Metropolitan Areas.* Santa Monica, CA: Milken Institute.

Drescher, Denise, April 13, 1998. *Research Parks in the United States: A Literature Review.* For PLAN 261 Department of City and Regional Planning, UNC-Chapel Hill. (www.aurrp.org).

Durso, Thomas W. 1996. "Home-Grown R&D." *The Scientist* 10(14):1.

European Commission, 1995. *Research and Technology: The Fourth Framework Programme (1994-1998).* Brussels, Belgium.

Fallows, James. 1994. *Looking at the Sun: The Rise of the New East Asian Economic and Political System.* New York: Pantheon Books.

Feller, Irwin. 1990. "Universities as Engines of R&D-Based Economic Growth: They Think They Can." *Research Policy* 19(4):335-48.

Florax, Raymond, and H. Folmer. 1989 (November). *Regional Economic Effects of Universities: The Impact of Knowledge Production on Investments of Industry.* Paper presented at the 1989 Meetings of the Regional Science Association. Santa Barbara, CA.

Florida, Richard. 1999. "The Role of the University: Leveraging Talent, Not Technology." *Issues in Science and Technology.* 16(1).

Franco, Michael R. 1985. *Key Success Factors for University-Affiliated Research Parks.* Ph.D. dissertation. University of Rochester.

Garbarine, Rachelle. Nov. 23, 1997. "Newark's Science Park Takes Another Step Forward." *The New York Times.*

Gibb, J. M., ed. 1985. *Science Parks and Innovation Centers: Their Economic and Social Impact.* Amsterdam: Elsevier Science Publications.

Glasmeier, A. 1987. "Factors Governing the Development of High-Tech Industry Agglomerations: A Tale of Three Cities." *Regional Studies* 22(4): 287-301.

Glasmeier, A. 1990. *The Making of High Tech Regions.* Princeton: Princeton University Press.

Goldberg, Carey. Oct. 8, 1999. "Across the U.S., Universities Are Fueling High-Tech Economic Booms." *The New York Times.*

Goldstein, Harvey A., and Michael I. Luger. Spring 1989. "Research Parks: Do They Stimulate Regional Economic Development?" *Economic Development Commentary* 13:3-9.

Goldstein, Harvey A., and Michael I. Luger. 1990. *Universities and Regional Economic Development in the 1990s.* Paper presented at the annual conference of the Association for Policy Analysis and Management, San Francisco, October 19, 1990.

Grindley, Peter, David Mowery, and Brian Silverman. 1996. "SEMATECH and Collaborative Research: Lessons in the Design of High-Technology Consortia." *Journal of Policy Analysis and Management,* 13(4).

Ham, R. M. and D. C. Mowery. 1995. "Improving Industry-Government Cooperative R&D." *Issues in Science and Technology.* 11(4).

Hamilton, Alexander. 1791. *Report on Manufactures.*

Kenney, Martin, ed. 2000. *Understanding Silicon Valley: The Anatomy of an Entrepreneurial Region.* Stanford: Stanford University Press.

Larson, Charles F. 2000. "The Boom in Industry Research." In *Issues in Science and Technology* 16(4).

Lebow, Irwin. 1995. *Information Highways & Byways: From the Telegraph to the 21st Century.* New York: Institute of Electrical and Electronics Engineers.

Levitt, Rachelle, ed., 1987. *The University/Real Estate Connection: Research Parks and Other Ventures.* Washington, D.C.: Urban Land Institute.

Link, Albert N. 1995. *A Generosity of Spirit: The Early History of the Research Triangle Park.* Durham: Duke University Press.

Link, Albert N. and Maryann P. Feldman, eds. 2001. *Innovation Policy in the Knowledge-based Economy.* Boston: Kluwer Academic Publishers.

Luebke, Paul, Stephen Peters, and John Wilson. 1985. "The Political Economy of Microelectronics in North Carolina," pp. 1310-28. *High Hopes for High Tech: The Microelectronics Industry in North Carolina.* Dale Whittington, ed. Chapel Hill: University of North Carolina Press.

Luger, Michael I. 1987. "The States and Industrial Development: Program Mix and Policy Effectiveness," pp. 29-64. *Perspectives on Local Public Finance and Public Policy.* John M. Quigley, ed. Greenwich, CT: JAI Press.

Luger, Michael I., and Harvey A. Goldstein. 1991. *Technology in the Garden: Research Parks & Regional Economic Development.* Chapel Hill: University of North Carolina Press.

Luger, Michael, Deog Song Oh, and David Gibson, eds. 1998. *Technopolis as Regional Development Policy.* World Technolopolis Association.

Malecki, E. J. 1991. *Technology and Economic Development.* New York: John Wiley.

Markusen, Ann, Peter Hall, and Amy Glasmeier. 1986. *High Tech America: The What, How, Where, and Why of the Sunrise Industries.* Boston: Allen and Unwin.

Miller, Roger, and Marcel Cote. 1987. *Growing the Next Silicon Valley: A Guide for Successful Regional Planning.* Toronto: D.C. Heath and Company.

Monck, C.S.P., et al. 1988. *Science Parks and the Growth of High Technology Firms.* London: Croom Helm.

Mowery, David C. 1998. "Collaborative R&D: How Effective is It?" *Issues in Science and Technology.* 15(1).

National Research Council. 1986. *The Positive Sum Strategy: Harnessing Technology for Economic Growth.* Ralph Landau and Nathan Rosenberg, eds. Washington, D.C.: National Academy Press.

National Research Council. 1995. *Allocating Federal Funds for Science and Technology.* Washington, D.C.: National Academy Press.

National Research Council. 1996. *Conflict and Cooperation in National Competition for High-Technology Industry.* Washington, D.C.: National Academy Press.

National Research Council. 1996. *Improving America's Schools: The Role of Incentives.* Eric A. Hanushek and Dale W. Jorgenson, eds. Washington, D.C.: National Academy Press.

National Research Council. 1999. *The Advanced Technology Program: Challenges and Opportunities.* Charles W. Wessner, ed. Washington, D.C.: National Academy Press.

National Research Council. 1999. *Harnessing Science and Technology for America's Economic Future.* Washington, D.C.: National Academy Press.

National Research Council. 1999. *Industry-Laboratory Partnerships: A Review of the Sandia Science and Technology Park Initiative.* Charles W. Wessner, ed. Washington, D.C.: National Academy Press.

National Research Council. 1999. *The Small Business Innovation Research Program: Challenges and Opportunities.* Charles W. Wessner, ed. Washington, D.C.: National Academy Press.

National Research Council. 2000. *The Small Business Innovation Research Program: An Assessment of the Department of Defense Fast Track Initiative.* Charles W. Wessner, ed. Washington, D.C.: National Academy Press.

National Research Council. Forthcoming. *The Advanced Technology Program: Assessing Outcomes.* Charles W. Wessner, ed. Washington, D.C.: National Academy Press.

National Research Council. Forthcoming. *Government-Industry Partnerships in Biotechnology and Information Technologies: New Needs and New Opportunities.* Charles W. Wessner, ed. Washington, D.C.: National Academy Press.

National Research Council. Forthcoming. *Regional and National Programs to Support the Development of the Semiconductor Industry.* Charles W. Wessner, ed. Washington, D.C.: National Academy Press.

Nelson, Richard R. 1986. "Institutions Supporting Technical Advances in Industry." *American Economic Review, Papers and Proceedings* 76(2):188.

Nelson, Richard R., ed. 1992. *Government and Technical Progress: A Cross-Industry Analysis.* New York: Pergamon Press.

Nelson, Richard R. 1998. *Technical Advance and Economic Growth.* Paper prepared for the NRC Forum on Harnessing Science and Technology for America's Economic Future, Washington, D.C., February. (www2.nas.edu/harness/21b6.html).

Okimoto, Daniel I. 1989. *Between MITI and the Market: Japanese Industrial Policy for High Technology.* Stanford: Stanford University Press.

Porter, Michael E. 1985. *Competitive Advantage: Creating and Sustaining Superior Performance.* New York: The Free Press, A Division of Macmillan, Inc.

Romer, Paul. 1990. "Endogenous technological change." *Journal of Political Economy* 98:71-102.

Rosenberg, Nathan, Ralph Landau, and David C. Mowery, eds. 1992. *Technology and the Wealth of Nations.* Stanford: Stanford University Press.

Saxenian, Annalee. 1994. *Regional Advantage: Culture and Competition in Silicon Valley and Route 128.* Cambridge, MA: Harvard University Press.

Sternberg, Rolf. 1990. "The Impact of Innovation Centres on Small Technology-Based Firms." *Small Business Economics.* 2(2):105-118.

Sternberg, Rolf. 1996. "Technology Policies and the Growth of Regions." *Small Business Economics.* 8(2):75-86.

Stokes, Donald. 1997. *Pasteur's Quadrant: Basic Science and Technological Innovation.* Washington, D.C.: Brookings Institute Press.

The Services Group. 1999. *A Feasibility Study for the Khadouri Technology Development Center.* Arlington, VA: U.S. Agency for International Development.

U.S. Department of Energy, 1995. *Task Force on Alternative Futures for the Department of Energy National Laboratories.* Washington, D.C.: U.S. Department of Energy.

von Hippel, Eric. 1988. *The Sources of Innovation.* New York: Oxford University Press.